S0-APP-252

22 Advances in Polymer Science

Fortschritte der Hochpolymeren-Forschung

With 77 Figures

Springer-Verlag
Berlin Heidelberg New York 1977

Editors

ISBN 3-540-07942-4 Springer-Verlag Berlin Heidelberg New York
ISBN 0-387-07942-4 Springer-Verlag New York Heidelberg Berlin

Library of Congress Catalog Card Number 61-642

Typesetting and printing: Schwetzinger Verlagsdruckerei. Bookbinding: Brühlsche Universitätsdruckerei, Gießen.

Contents

Relaxation and Viscoelastic Properties of Heterogeneous Polymeric Compositions

Yuri S. Lipatov
Institute of Macromolecular Chemistry, Ukrainian Academy of Sciences, 252160, Kiev, USSR

Table of Contents

1. General Definitions and Principle Causes of Change of Properties in Thin Polymeric Layers on Solid Surfaces

Heterogeneous polymeric compositions include most of commercially important polymeric materials, such as reinforced plastics (including fiber glass reinforced and carbon plastics), thermoplastic and thermosetting polymers filled with dispersed fillers. All these systems are two-phase systems in the least of which heterogeneity is inherent in the very principle of production of these materials. Polymeric mixtures are noted for a two-phase structure in which both the phases are continuous, and for this reason it is impossible to tell which of the polymers is the dispersion medium and which the dispersed phase. In contrast to mixtures, polymers, filled with polymeric fillers, comprise systems which are characterized by a specific nature of distribution of one component in the other. Despite such demarcation, the physico-chemical properties of the latter two types of systems are similar. In considering the viscoelastic properties of heterogeneous systems they can de divided into two main groups: (*1*) in which the dispersed phase inclusions (fiber glass, inorganic filler, etc.) are virtually incompressible, as compared with the polymer matrix and (*2*) in which both the components possess approximately the same compressibility. These two types of systems also differ in the structure of the transition region between the two phases. The first type, due to the adsorption interaction of the polymer with the solid surface, is characterized by the appearance of a surface layer of varying thickness whose properties differ from properties within the bulk (*1, 2*).

The thickness of this layer depends on the polymer cohesion energy, free surface energy of the solid, and on the flexibility of the polymer chains. The second type of the systems is characterized by the appearance of a transition region in which the structure and properties of the surface layers of both phases undergo changes because of the interaction of the components (*3*), limited compatability (*4*), and mutual diffusion. The difference in the structure of the two types can be schematically represented as in Fig. 1.

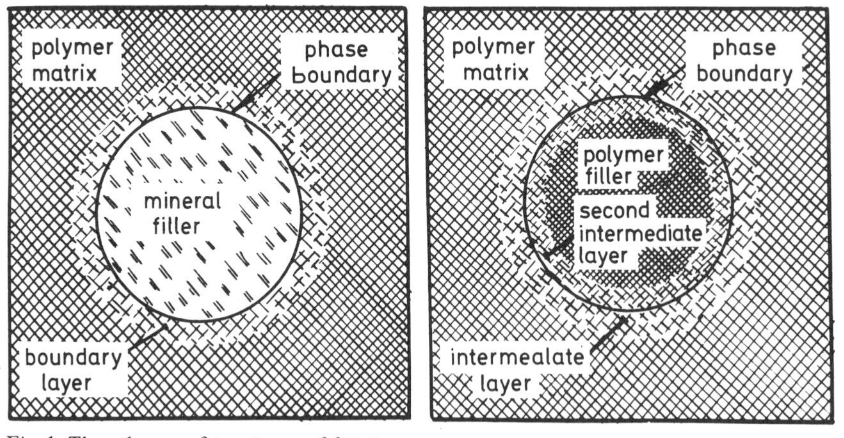

Fig. 1. The schemes of two types of filled systems

In both the systems, as a result of adsorption and adhesion interaction between the components at the interface, the boundary layers undergo changes, as compared with the respective properties within the bulk. This brings about the appearance of additional levels of microheterogeneity (*5*).

These include:

1) Molecular heterogeneity, manifested by change in the interphase of such physical characteristics which are determined by the macromolecular structure of the polymer chains (thermodynamic properties, molecular mobility, density of packing, free volume, etc.).
2) Structural microheterogeneity, which is determined by changes in the mutual disposition of molecules in the surface and transition layers at various distances from the phase boundary and which characterizes the short-range order in amorphous – and the degree of crystallinity in crystalline polymers.
3) Microheterogeneity on the supermolecular level determined by differences in the type and character of formation and packing of supermolecular structures both in the surface layer and within the bulk.
4) Chemical microheterogeneity caused by the effect of the interface on the reaction of formation of polymeric molecules. This type of microheterogeneity may be an additional cause of the previous three types of heterogeneity.

The above factors have a complex effect on relaxation and viscoelastic properties of heterogenic polymeric materials.

2. Some Concepts Concerning Mechanical Properties of Surface Layers of Polymers in Homogeneous Systems

Mechanical properties of thin polymeric films and the influence of properties of surface layers on viscoelastic properties have been given comparatively little consideration. In a number of investigations (*6, 7*) the mechanical properties of thin films have been studied in terms of their thickness. It has been shown (*6*) that the elastic limit rises and the ultimate elongation drops with the decrease of the film thickness. No changes have been found in another study (*7*). However, on the basis of the research (*8, 9*) it can be expected that the mechanical characteristics of thin films, associated with relaxation processes, should differ from the properties of block specimens. The difference will be more appreciable, the greater the ratio of the area of the film surface to the volume, when the contribution of the surface layers into the general mechanical properties increases. It is evident from (*8, 9*) that polymer molecules on the surface and close to it cannot have the same conformations as those within the polymer bulk. For this reason the feasibility of changes in the conformations of polymer molecules during deformations in the surface layer reduces, which is equivalent to the increase of the average relaxation time or to the shift of the relaxation spectrum toward larger periods. This must lead to an increase in the stiffness of the polymer surface layers.

In the work of (*10*) an evaluation has been made of the thickness at which the influence of the surface layer on the general mechanical properties becomes appreciable.

Using polymethylmethacrylate (PMMA) and nitrocellulose films we experimentally determined the modulus of elasticity E and tangent of the angle of mechanical losses tg δ, employing the method of forced flectural oscillations of cantilevered films in accordance with a technique specially worked out for this purpose (*10*). The curves of mechanical losses versus frequency for the thinner films are steeper and are shifted toward greater frequences, as should have been expected. Those data have been interpreted as evidence of the narrowing of relaxation time spectrum.

Figure 2 (*10*) illustrates the relationship between the modulus of elasticity and the thickness, while the curve of correlation between the surface area of the specimen S and volume V versus the film thickness is given for the sake of comparison. An appreciable increase of the modulus is observed just in the region of those thicknesses in which the correlation rises. This enables us to assert that the noted charge in the mechanical properties of thin films is caused by the effect of the polymer surface

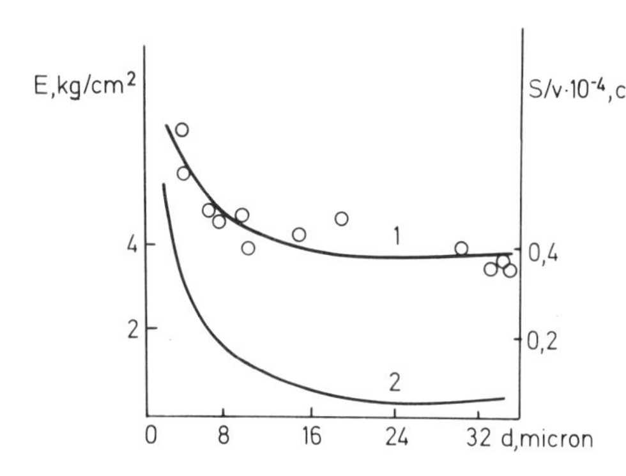

Fig. 2. Dependence of the elasticity moduli E (1) and the ratio of the surface area S to specimen volume V (2) on the film thickness d

layer. The findings indicate that surface properties of the polymer differ from those within the bulk because of the shift of the relaxation spectrum of the surface layer toward greater periods of time. It means that the polymer surface is stiffer than the bulk.

A tentative estimation of the thickness of the surface layer whose properties differ from those within the bulk was made in (*11*), considering a thin film as a three-layer plate in which the external layers are formed by the polymer surface layer, as the internal layer is formed by its bulk portion. The following equations have been proposed for this case:

$$E_s = \frac{48\pi^2 \, l^4 \nu^2 \, [\rho_0 \, (d-d_s) + 2\delta \, s \, d_s]}{(1.875)^4 \, [d^3 - (d - 2\, d_s)^3} - E_0 \frac{(d - 2\, d_s)^3}{[d - (d - 2\, d_s)^3]} \tag{1}$$

$$\mathrm{tg}\, \delta_s = \left[1 + \frac{E_0 \, (d - 2\, d_s)^3}{Es \, [d^3 - (d - 2\, d_s)^3]}\right] \frac{\Delta\nu}{\nu} - \frac{E_0 \, (d - 2\, d_s)^3}{E_s \, [d^3 - (d - 2\, d_s)^3]} \, \mathrm{tg}\, \delta_0 \tag{2}$$

in which E_s and tg δ_s = dynamic modulus and mechanical losses of the surface layer

l = length of the rod

ν = resonant frequency of oscillations

$\Delta\nu$ = width of the resonance curve
ρ_0 = density of polymer within the solid
d_s = thickness of surface layer
E_0 and tg δ_0 = modulus and losses of polymer within the bulk.

Some results of the calculations are given in Table 1 (*11*).

Theoretically the mechanical properties of surface layers were explained in the work (*12*). The mechanical properties of surface layers were estimated with the aid of transient processes of supersonic oscillations of the surface. The transient process of mechanical oscillations of a system is characterized by the process reaction to the pertubation factor. This reaction is definitely determined by the mechanical properties, and for a single action is described by a transient function $F(t)$, *i.e.* by the time-dependent characteristic of the mechanical properties. Using a specially elaborated technique, transient processes in surface layers were investigated on electric circuits. For subsequent transition to mechanical models use has been made of electric-to-mechanical analogs. This technique has enabled mechanical models of the surface layer based on direct experimental findings to be obtained for the first time. Figure 3 illustrates the mechanical models, and Table 2, their parameters obtained from the electric circuits. As is evident, the mechanical models of the surface layer differ one from another both in the disposition of the elements and in the magnitude of the mechanical constants. It is significant that the parameters of the models differ in terms of the frequency of the action, which could have been expected in line with the considerations given at the beginning of the present sectiom

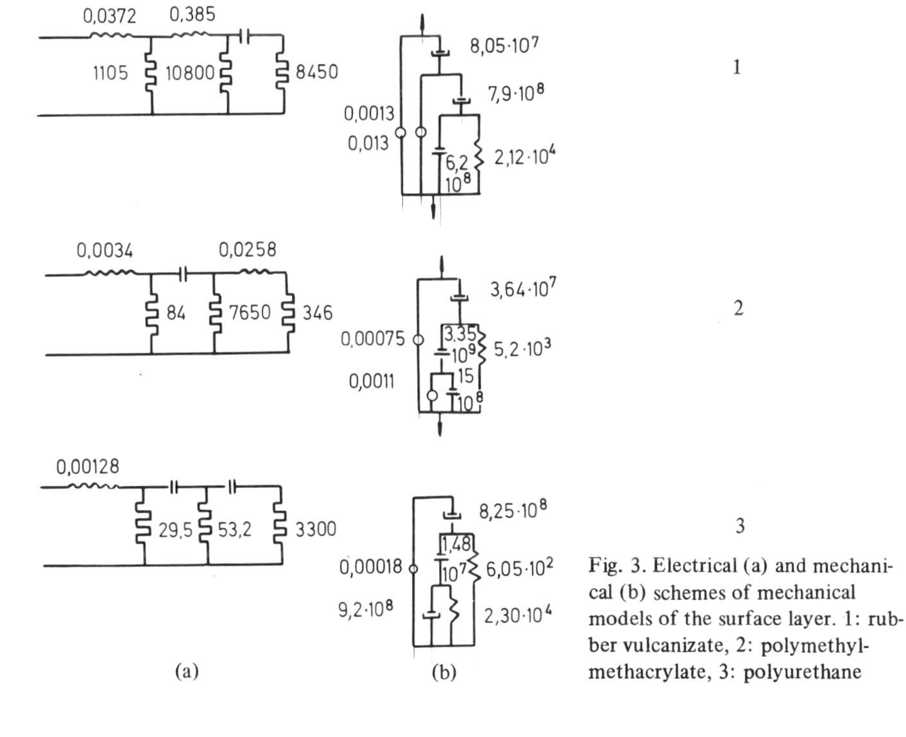

Fig. 3. Electrical (a) and mechanical (b) schemes of mechanical models of the surface layer. 1: rubber vulcanizate, 2: polymethylmethacrylate, 3: polyurethane

Table 1. Mechanical properties of polymers in bulk (E_0 tg δ_0) and at the surface (E_s tg δ_s) (frequency 500 c/sec)

Polymer	E_0 kg/cm^2 1.10^{-4}	tg δ_0	d_s mcm	E_s kg/cm^2 1.10^{-4}	tg δ_s
Polymethyl-methacrylate	3.5 ± 0.15	0.7 ± 0.1	0.6 ± 0.2	9 ± 2	0.145 ± 0.02
Nitrocellulose	2 ± 0.11	0.6 ± 6.1	0.35 ± 0.12	7 ± 1.5	0.13 ± 0.015

Table 2. Values of parameters of mechanical moduls, obtained from the electric circuits

Parameter	Vulcanized Rubber	Polyurethane	Polymethylmethacrylate	
	320 kHz	320 kHz	320 kHz	126 kHz
Elasticity modulus kg/mm^2 E_1	$2.12 \cdot 10^4$	$605 \cdot 10^2$	$5.2 \cdot 10^3$	$3.99 \cdot 10^3$
The same E_2	–	$2.30 \cdot 10^4$	–	–
Vibrating mass, g m_i	0.00129	0.00018	0.00075	0.00368
The same m_2	0.0134	–	0.0011	0.0185
Viscosity, phases	$8.05 \cdot 10^7$	$8.25 \cdot 10^6$	$3.64 \cdot 10^7$	$3.46 \cdot 10^8$
The same	$7.9 \cdot 10^8$	$1.48 \cdot 10^7$	$3.35 \cdot 10^9$	$3.89 \cdot 10^8$
The same	$6.2 \cdot 10^8$	$9.20 \cdot 10^8$	$1.5 \cdot 10^8$	$1.11 \cdot 10^6$
Electromechanical equivalent	$3.48 \cdot 10^{-2}$ $7.3 \cdot 10^{-4}$	0.142 $2.78 \cdot 10^{-3}$	0.22 $4.32 \cdot 10^{-3}$	0.44 $1.11 \cdot 10^{-2}$

Thus the obtained models, resulting from experiments and not from phenomenological concepts, also make it possible to trace the change in the relaxation behavior, when passing from one frequency to another.

The same method of surface ultrasonic waves can also be used for determining that thickness of the surface layer in which the properties differ from the properties within the bulk (*13*). The values obtained for polymers of various chemical nature are within 200 and 700 μ, depending on the thickness of the surface layers of the polymers in heterogeneous filled systems (*14*). It follows from the theoretically obtained equations that the thickness of the layer derived from the data on the propagation of surface ultrasonic waves, depends both on the mechanical properties of the bulk and surface layer and on the frequency. The difference in the modulus of elasticity of the bulk and surface layer are associated with the surface tension forces. The frequency-dependence of the thickness is determined by the types of molecular motions involved in the process in accordance with the mechanical models indicated above.

3. Effect of the Surface on the Transition Temperatures in Heterogeneous Polymeric Systems

As is known, the temperature of transition from the highly elastic to the glass-like state is one of the most important characteristics of the polymer viscoelastic properties, since it determines the position of the transient region.

In heterogeneous systems, containing a polymeric or inorganic filler, as the self-contained phase, the interphase interactions can change the molecular mobility, thus affecting the position of the transition temperatures.

Up to the present a large number of experimental data have been published indicating a change of T_g caused by the effect of the solid surface. These data have been obtained by the use of various methods (dilatometry, dynamic and mechanical properties, thermal capacity, NMR, and others). Since each of the methods has its own limitations and describes mainly one type of molecular motions (*15*), the results obtained by comparing the data received by different methods do not always correlate. Figure 4. (*16*) illustrates changes of T_g in PMMA caused by the introduction

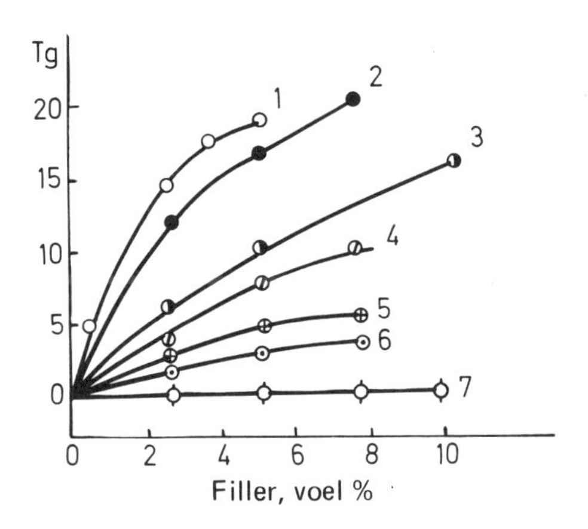

Fig. 4. Dependence of the change in the glass temperature for PMMA (1–5) and PS (6, 7) on aerosil concentration for different methods. 1: calorimetry, 2: dilatometry, 3: dynamical properties, 5: NMR, 4,6-dielectrical relaxation, 7: calorimetry, dilatometry, mechanical

of Aerosil, determined by different methods. As can be seen, though the general character of the change of T_g with the increase of the filler concentration is preserved, the magnitude of change of T_g is unequal. Tg is a the maximum when methods with the large time-dependent scale of the test are employed (dilatometry, calorimetry). The increment of Tg decreases as the frequency increases.

Experimental data concerning the effect of fillers on Tg of elastomers differ considerably.

In a number of investigations (*17–21*) a certain increase of T_g was noted, while in other works (*22–25*) such effect was not indicated. In certain cases they noted even a decrease of Tg (introduction of 10% chalk to PMMA decreases T_g by $\approx$ 10% (*26, 27*). The combined effect of fillers and plasticizers on the increase of T_g is discussed in (*28*).

However, in most cases, particularly in considering Tg of filled thermoplastic polymers, as rule, a substantial increase of T_g is indicated (*1, 29–31*). Some regularities of the change of T_g will be discussed below (*30–34*). Some data give evidence to the fact that an increase of T_g, the degree of volumetric filling being the same, depends on the total surface of the filler.

Since Tg is not a thermodynamic point, though its position is associated with the beginning of the manifestation of the mobility of chains, the findings indicate that the formation of physical bonds between polymeric molecules and the surface, and the conformation changes in the surface layer, lead to a change in the relaxation behaviour of the polymer. An increase of Tg indicates an appreciable restriction in the mobility of chains, a restriction which is equivalent to a decrease of their elasticity, caused by the formation of additional bonds or a change in the conformation of the chains. The restriction of mobility, which manifests itself in the increase of T_g, will be, of course, dependent on the number of chain segments, which participate in the interaction with the surface. Therefore the increase in this number will lead to the increase in both the molecular mobility and T_g. This was indicated in the work (*35*) for the case of studying differences in the increase of T_g for fillers of polystyrene (PS) and PMMA films obtained from solutions in thermodynamically different solvents. An important conclusion which follows from the increase of Tg in the presence of fillers is that the change in the properties of the polymer on the surface, the change leading to the increase of T_g, concerns not only the polymer layers which are in direct contact with the surface, but also the more remote layers. Indeed, if the change of mobility had concerned only the layers of macromolecules which are in direct contact with the surface, charges of such value as T_g, which characterizes the behaviour of a large number of molecules, would not be detected. Glass transition is actually a cooperative process, and for this reason, limitations of mobility imposed by a hard surface apply both to the molecules in direct contact with the surface and to molecules bonded into aggregates (*2*), and to other supermolecular structures. Thus the increase of T_g in the presence of a filler is due to the interaction with the surface of both individual (isolated) macromolecules and their aggregates (*1–3, 6*). The restriction of mobility of one chain interacting with the surface will lead to the restriction in mobility of other chains bonded into aggregates. However, in filling, we could not observe separate transition temperatures for the surface layers and for the polymer bulk not affected by the surface influence. The possible explanation is that with a low content of the filler the feasible difference in the transition temperatures in the surface layer and inside the substance is not great, and both transitions merge into one, shifted toward greater temperatures. With a greater content of fillers, the surface affects most of the polymer, and again only one transition is observed. Such position of transitions is highly probable, if one takes into consideration (see below) that in filling an expansion of the relaxation spectrum takes place, as well as a spread of the temperature interval between the beginning and end of the glass transition (which can conveniently be observed with the DTA technique (*37*). Separate transition temperatures associated with the existence of surface layers were observed only in a few experiments (*38–40*).

The increase of the glass transition temperatures, which is clearly observed during the determination of the jump in thermal capacity in glass transition, is a magnitude which makes it possible to calculate the effective thickness of the surface layer. The ΔC_p magnitude is regularly reduced with the increase of the content of the filler. The decrease of the magnitude is a definite indication that a certain part of macromolecules is excluded from the cooperative process of glass transition because of the interaction with the surface. This part of macromolecules forms the boundary layer proper of macromolecules close to the surface, where the mobility of macromolecules is considerably reduced. In this case the portion of excluded macromolecules f can be found proceeding from the relationship

$$f = 1 - \Delta Cp_f / \Delta Cp \tag{3}$$

in which ΔCp_f = thermal capacity jump for the filled and ΔCp = for the unfilled polymer (*41*). Then the thickness of the surface layer can be approximately evaluated from the equation

$$\left(\frac{\Delta r + r}{r}\right)^3 - 1 = f \frac{\varphi}{1 - \varphi} \tag{4}$$

in which r = diameter of the filler particle
φ = volumetric content of the filler
Δr = thickness of the boundary layer.

The f value is very convenient for the analysis of the dependence of T_g on the filler content.

Most experimental data indicate that T_g dependence on the filler content is non-linear (*42*), and in the form is similar to the well-known dependence of T_g on the polymer molecular weight.

It might seem that since the change of T_g is associated with the restriction of molecular mobility, it would have to manifest itself the more, the thinner the polymer layer between two particles of the filler, *i.e.* the greater its concentration. Since this is at variance with experimental findings, it becomes evident that T_g is mostly affected, not by the thickness of the layer but by other factors. Those were analyzed in the work of (*43*). It was found that similar to T_g, f, with the increase of the filler concentration, also tends, toward a limiting value. It appears that there is a linear dependence between Tg of an unfilled polymer and the fraction of polymer in the boundary layer, namely,

$$T_g = T_{g_0} + f \Delta T \tag{5}$$

where ΔT is the maximum increment of Tg for a system in which the polymer phase is affected by the solid surface ($f = 1$).

The values of f and ΔT in the above equation are dependent on the density of cohesive energy Ec and on the stiffness parameter of an individual macromolecule σ (*43*). The magnitude of f tends towards an increase with the increase of σ. For polymers with approximately equal σ the value of f changes antipathically to the density of cohesive energy of the polymer Ec. In a general form f may be represented as

$$f = f(S, \sigma, E_f/E_c) \quad (6)$$

in which S = filler surface area
E_f = surface energy of the filler

Indeed, the transition of macromolecules into the boundary layers is facilitated with the decrease of the decrease of the intermolecular interaction in the polymer and with the increase of the polymer-filler interaction energy. The value of ΔT increases sympatically to E_c. It follows that the changes of T_g caused by the filler surface are much dependent on the cohesion energy of the polymer. For polymer with low cohesion energies (*e.g.* rubbers) the transition to the surface layer state has little effect on their properties, ΔT is little and T_g changes insignificantly with a change of the filler concentration. In case of a strong intermolecular interaction, the presence of even a small portion of macromolecules will result in a substantial increase of Tg because of a large value of the parameter ΔT.

4. Volume Relaxation in Filled Systems

In accordance with the concepts developed in (*44*) the glass transition temperature is dependent on the mobility of both very small and large structural elements of the chains, even on the mobility of supermolecular structures. For this reason the glass transition temperature cannot be considered only as a temperature which determines the manifestation of segmental mobility.

Obviously, the presence of a loose temperature glass transition interval can be explained by the fact that various mechanisms participate in the process of glass transition. Of interest from this point of view is the examination of the effect of fillers on such relaxation processes in which sufficiently large structural elements take part. Those phenomena were studied (*33, 45–47*) by calculating the average relaxation time proceeding from the data on the isothermic contraction of the volume of various filled systems by the method proposed in (*48*).

The analysis of the data obtained for a number of systems, containing both inorganic and polymeric fillers, shows that the average time of relaxation of the process of isothermic contraction τ_{av} increases in the presence of hard particles. However, from the point of view of understanding the mechanism of the processes which occur in the interface, it is more correct to compare the relaxation time at corresponding temperatures equidistant from the glass transient temperature, rather than at equal temperatures.

As is seen in Fig. 5, which represents the relaxation time versus the difference between the temperature in the experiment and T_g, the relaxation time in such a comparison for the system examined in the presence of a solid surface, is smaller than for an unfilled polymer.

The increase of T_g in the surface layer explains the greater relaxation time in filled systems compared at equal temperatures. However, some of our research indicated a reduction of the density of packing of polymers in the surface layer, and an increase of the free volume (*1*). This effect reduces the average relaxation time in

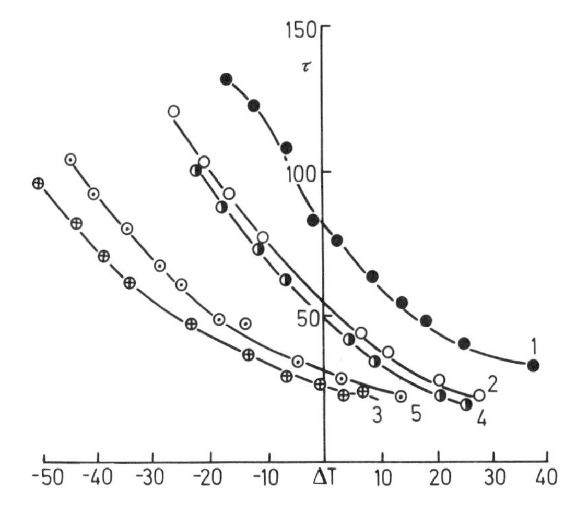

Fig. 5. Dependence of the average relaxation time of isothermical contraction on the difference between measurement and glass temperatures for different systems. 1: low molecular PMMA, 2: the same 5% of glass fiber, 3: the same + 30% of glass fiber, 4: the same + 5% of acrylonitryle fiber, 5: the same + 30% of acrylonitrile fiber

filled systems as compared with unfilled systems, provided comparison is performed at corresponding temperatures.

It is of interest that the phenomena discussed above are most clearly manifested for linear polymers. When passing over to cross-linked polymers (*e.g.* copolymers of styrene and divinylbenzene), the effect of the surface on the relaxation time becomes less perceptible as the network density increases. This is due to the fact that the reduction in the mobility of large segment of the chains caused by cross-linking excludes them from participation in the relaxation process, thus leveling the effect of the surface on their mobility.

The data on the temperature dependence of the average relaxation time have made it possible to calculate the activation energy E_a. A certain increase of E_a was noted with the increase of the filler content, *i.e.* with the reduction of the surface layer thickness. Above T_g Ea is virtually independent of the layer thickness. The activation energy below T_g becomes smaller, the higher the Tg, which results from the loose packing of macromolecules in the surface layer, and from the participation in the process of solely structural elements with a small relaxation time. The complex dependence of E_a on the content of solid phase within the region of temperatures below and above T_g is thus connected with the simultaneous influence on the relaxation process of such factors as loose packing, stiffening of chains in the surface layer because of a reduced number of possible conformations (*2, 8, 9*) and other factors. All the effects are reflected in the temperature dependence of the viscoelastic properties of filled polymeric systems.

The above discussion confirms that T_g of a filled polymer and relaxation processes in it are dependent in a complicated manner on the correlation of entropy and energy effects occurring in interaction with the solid surface. Along with it, the analysis of numerous data (see Section 5 below) indicates that the mobility change effects are dependent on the flexibility of the chains. It may be assumed that both the excessive stiffness of the chain and excessive flexibility will level the effect of the surface on the change of molecular mobility. Those processes will be most clearly manifested in chains of flexibility intermediate between the flexibility of rubber chains and of stiff-chain polymers (*i.e.* mainly for thermoplastics).

5. Nuclear Magnetic Resonance and Dielectric Relaxation

The analysis of dielectric relaxation in polymers enables the mobility of functional groups of chains and segments to be studied separately, since the regions of dielectric dispersion are manifested in different conditions for different structural elements of the chains (*49, 50*).

The method of nuclear magnetic resonance which has been widely developed for the last decade also makes it possible to estimate the character of molecular movements in polymers (*51*). This enables the above methods to be used for the study of relaxation processes in filled polymeric systems. We began such study for the first time in 1965 (*52*), proceeding from the concept that changes in transition temperatures in the presence of fillers reflect changes in the character of relaxation processes in thin layers on the surface of the filler. It could be supposed that the surface forces at the interface exert various influence on one or other relaxation mechanisms which participate in the general relaxation process. In a number of such studies we examined in detail the processes of dielectric relaxation in the surface layers of PS, PMMA, copolymer of styrene and methylmethacrylate, cellulose acetate, polyurethanes, as well as of some oligomers (*52–60*). We examined polymers differing in the character of functional groups and flexibility of chains and surfaces with high and low surface energies. Changes in the molecular mobility of polymeric chains can be characterized by the magnitude of the shift of the maximum of dielectric losses on the curve showing the dependence of tg δ the temperature. Figure 6 represents the temperature curves for PS and PMMA in the presence of aerosil and teflon powder. The content of the fillers taken with due account to the size of their particles is such that the thicknesses of the surface layers may be approximately equal. It can be seen in the drawing that with the reduction of the thickness of the surface layer one can observe a shift of the maximum of corresponding to the dipole-group relaxation process, toward lower temperatures, and of the maximum of the dipole-segmental process toward higher temperatures. This indicates a change in the average relaxation time of the respective processes in the surface layers. The temperature dependence for copolymer of styrene and methylmethacrylate also indicate a shift of the low-temperature maximum to the left, and of the high-temperature maximum to the right, due to the introduction of the filler.

The findings indicate that the presence of the polymer-solid interface has indeed a diverse effect on different relaxation mechanisms. The interaction of functional groups with the surface and the resultant restriction in the mobility of the chains and formation of a sparse space network should have led to an increase of the relaxation time and to a shift of the position of the maximum of dielectric losses toward higher temperatures. However, this has not turned out to be true. As a rule, the peak of losses, corresponding to the motion of side groups of the polymer which is in the interface, shifts toward lower temperatures, the degree of the shift being dependent on the nature of the polymer and thickness of the surface layer. This indicates an increase of mobility of polymer chains in the interface and is associated with a reduced density of polymer packing on the surface of solid particles. It is known from the theory of adsorption of polymers on solid surfaces (*2*) that only a portion of chain segments is directly bonded to the surface, whereas most of them remain free.

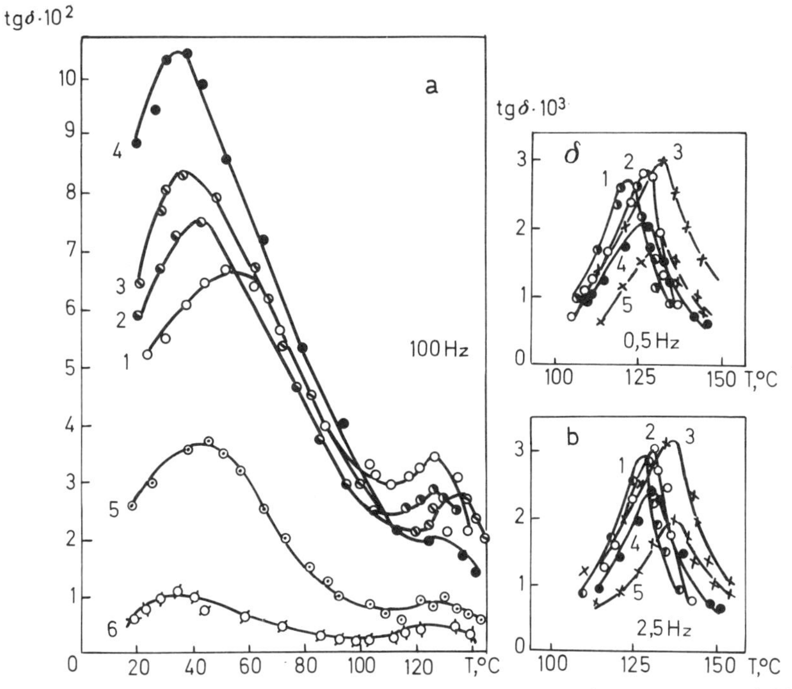

Fig. 6. Dependence of tg δ on temperature: a) PMMA 1: initial, 2–4: with aerosil 1,32; 12,5; 23,1%, 5: with 49,2% teflon, 6: with 75% of teflon. b) PST 1: initial, 2: 1,32, 3: 23.1% of aerosil, 4 and 5: 49,2 and 75,0% of teflon

If we take into account that not only individual (isolated) molecules but also molecular aggregates can bind with the surface, it becomes clear that most of the functional groups on the surface are not directly bonded to the latter (*61*). This, combined with a reduced density of the packing, explains greater freedom of molecular motion.

The mobility of large structural elements of chain segments diminishes in the interface with the solid. Consequently, a specific "scattering" of maximums of dielectric losses occurs in the surface layers, which indicates an expansion of the relaxation time spectra in the boundary layers as compared with the bulk. Along with it, the increase of T_g for the surface layers which points out an increase of the average relaxation time, shows this effect is dependent mostly on relaxation time of the greatest structural elements with low mobility. Changes of properties at the interface can also be characterized by the width of the relaxation time spectrum which is described by the Cole-Cole parameter of relaxation time distribution (*62*). Figure 7 illustrates circular diagrams of generalized dielectric permeability which we calculated for copolymer methylmethacrylate with styrene. These were used for computing the parameter of distribution of the relaxation time α (*63*). Figure 8 shows dependences $\alpha = f(T)$ for copolymers. It is seen that the parameter of relaxation time distribution decreases with the decrease of the thickness of the surface layer in the region of dipole-group losses and increases in the region of high-temperature dispersion. This also gives evidence for the expansion of the relaxation spectrum. It follows from

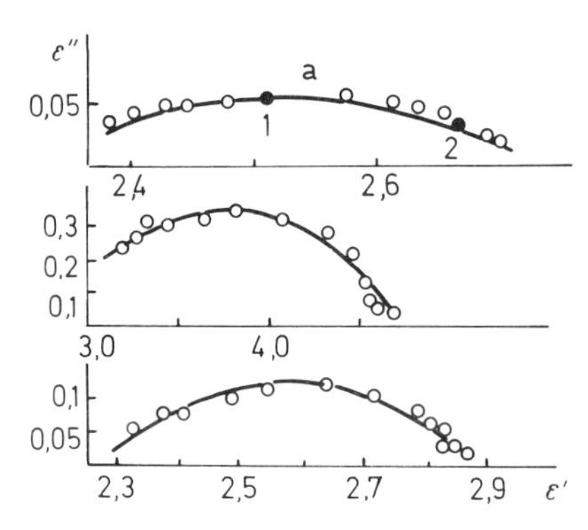

Fig. 7. Cole-Cole circular diagrams of generalized dielectric permeability for PMMA-PS copolymer. a) initial polymer, 40°, b) with 24,9% of aerosil (142,5°), c) with 75% of fluoroplast (134,5°)

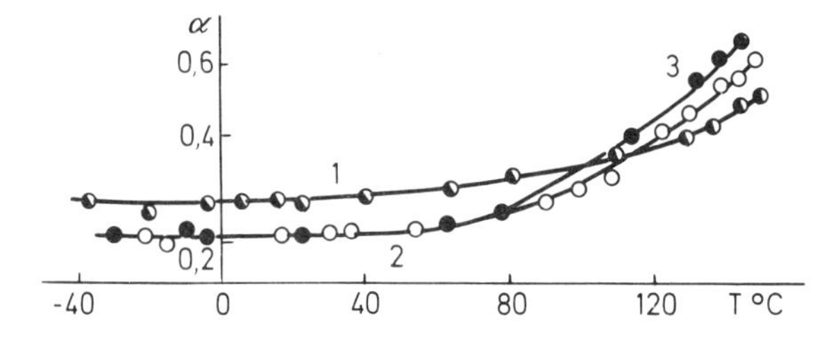

Fig. 8. Temperature dependence of the parameter of relaxation time distribution (α) for copolymer: 1: initial, 2: with 24,9% of aerosil, 3: with 75% of fluoroplast

(*64, 65*) that that the decrease of the parameter α (region of dipole-group process) really indicates a smaller density of packing of the polymer in the surface layers and the formation of a large number of relaxants, as well as the increase of the mobility of functional groups of chains and of small kinetic units unbound to the surface, while for the region of the dipole-segmental process the increase of the distribution parameter also indicates a hindering of the movement of the segments at the interface (narrowing of the relaxation time spectrum).

The data obtained in our experiments for fillers of various surface energy are essential in that the observed changes in the mobility of side groups and segments (dipole-group and dipole-segmental losses) are not very dependent on the nature of the filler surface. A very important conclusion follows that changes in the mobility are mainly affected by the geometric restriction of the number of possible conformations of macromolecules near the particle surface, *i.e.* by the entropic factor. Specifically, these restrictions inhibit such dense packing of molecules which would have occurred in the bulk. This principle is confirmed by the results of the studies of molecular mobility at the interfaces of the stiff-chain polymer cellulose acetate. The temperature curves for the cellulose acetate in the bulk and on the surface of modified and non-modified aerosile show that in the stiff-chain polymer, the conformational set of chains, which is greatly limited as compared with flexible molecules, cannot change near the interface as greatly as in the case of flexible molecules; and in this case the effects of changes of flexibility of chains do not manifest themselves. Similar results were obtained for the surface layers of acrylate-epoxy-styrene compositions in which the effects of the surface diminished with the increase of stiffness of chains (*67*). However, it can be controlled only in the case of a change in the chemical nature, and this does not allow the role of each factor to be clearly discriminated.

The increase in the flexibility of chains in the surface layer leads to a reduced effect of the surface on the molecular mobility. This was shown in experiments on the use of fillers with molecules of various lengths grafted on their surface (*68*). Molecular mobility was studied in epoxy resin in the presence of aerosil modified by alcohols from butyl to tetradecyl. It was established that a relaxation process characterizing the grafted molecular mobility itself is manifested with the increase of the length of the grafted molecule. When using such modified fillers in which the grafted molecules can play the role of plasticizers for polymer molecules with an increase in the length of the grafted molecules, an ever smaller shift is observed of the maximum of dipole-segmental losses toward higher temperatures, and a smaller shift of the maximum of dipole-group losses toward lower temperatures. This bears evidence to a smaller stiffening of the polymer molecules in the boundary layer and a smaller loosening of packing of polymer chains in the surface layer in the presence of the modified additions. Thus, the modification of the filler by organic molecules chemically bonded to the filler surface, molecules of sufficient length to enable them to manifest their own molecular mobility in the surface layer, reduces the effect of the surface on the temperature shifts of the relaxation processes in the boundary layers of the polymers. Hence, such modification brings about a realization of a more balanced state of equilibrium of the polymer in the boundary layer, and may be used for the control of the characteristics of filled polymers.

The pattern of dielectric relaxation undergoes no essential changes in passing over from filled linear to filled three-dimensional polymers. Thus, in the study of the dielectric relaxation in cross-linked polyurethanes (*55*) of various nature is was shown that the maximum of the dipole-group process with aerosil as the filler, shifts toward lower temperatures, while the high-temperature process shifts toward higher temperatures. However, for polyurethanes a third peak is indicated on the dielectric relaxation curves at temperatures above the temperature which corresponds to the main maximum of the linear polymer. The temperature increase contributes to the rupture of some physical bonds in the lattice, thus releasing larger structural elements whose mobility is manifested in the temperature region of the third peak. This (third) maximum, upon the introduction of the filler, also shifts toward higher temperatures, which testifies to the formation of an additional number of physical bonds with the surface of the filler particles, *i.e.* to the increase of the efficient density of the network. Hence, the analysis of the circular diagrams for three-dimensional polymers gives evidence for the existence of at least three relaxation processes.

All the findings indicate that the mobility change processes affect not only the adsorption layers; they extend at greater distances from the surface (if it were not so, the effects of the shifts of loss peaks would be unnoticable).

Information on the change of molecular mobility in boundary layers of polymers can also be obtained by the use of nuclear magnetic resonance technique. Numerous available date (*69*) indicate that the research of the relaxation processes in polymers carried out by the use of the dielectric relaxation or nuclear magnetic resonance methods yield similar results. In a number of works on the objects discussed above, we studied the spin-lattice relaxation of protons T_1 in polymers and oligomers which are on the surface of filler particles (*53–56*). As an example, consider the data on the temperature dependence of the time of the spin-lattice relaxation of protons for polystyrene and for specimens containing aerosil and polyfluoroethylene (Fig. 9).

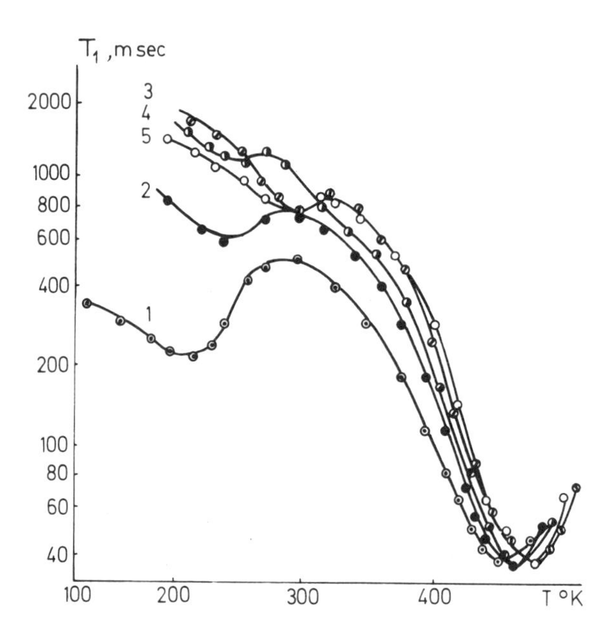

Fig. 9. Dependence of spin-lattice relaxation time T_1 on temperature for polystyrene: 1: initial, 2: with 1,3%, 3: with 23% of aerosil, 4 and 5: with 49,2 and 75% of fluoroplast

There are two relaxation regions, viz. high-temperature and low-temperature. For the high-temperature region the minimum of T_1 is shifted toward higher temperatures as the thickness of the surface layer diminishes, and the shift reaches 20°. At the same time, the low-temperature process is shifted toward lower temperatures. For a number of the analyzed systems general regularities were established similar to those explained above for processes of dielectric relaxation, which gives evidence for the common character of the mechanisms of changes of the relaxation behavior of polymers in boundary layers. Specifically, the findings support the fact that the phenomena observed during the study of dielectric relaxation do not result from the Maxwell-Vagner effect of heterogeneity of medium which is characteristic of objects with conductive and non-conductive regions (*70*). The study of filled polymers by the method of dielectric relaxation and nuclear magnetic resonance discussed in a number of works by Solomko (*71, 72*) confirms the general concepts put forth above.

Now consider the problem of the energy of activation E_a of relaxation processes in surface layers. The magnitudes of E_a can be calculated separately for the high- and low-temperature processes in the case of dielectric relaxation from the temperature dependence data $\ln f_m = f(T)$ in which f_m = frequency corresponding to the maximum of dielectric losses, and from the temperature dependence of the correlation time in the case of the use of the nuclear magnetic resonance technique. It should be noted that the values determined by this method for the reason of complexity of the processes which occur at a time, do not represent the true activation energy but the temperature coefficient of the relaxation time (*69*).

The principal regularities established in determining the E_a magnitudes indicate that E_a of high-temperature processes found both by the NMR technique and from dielectric relaxation, diminish with the reduction of the surface layer, while E_a of the high-temperature process changes little or even diminishes slightly. Some data illustrating the latter statement are given in Table 3. The findings on the activation

Table 3. Activation energies of relaxation in thin layers

Aerosil content, %	Fluoroplast, content, %	NMP activation energy, kcal/mol			Dielectric activation energy, kcal/mol		
		PMMA	PS	Copolymer MMA-styrene	PMMA	PS	Copolymer MMA-styrene
Group motion							
0.0	0.0	1.75	–	2.05	23.7	–	14.9
8.83	–	–	–	–1.75	–	–	12.6
1.32	–	1.36	–	–	18.5	–	–
23.08	–	1.18	–	–	15.4	–	–
24.9	–	–	–	1.47	–	–	10.7
–	26.5	–	–	1.72	–	–	12.6
–	49.2	1.48	–	–	20.0	–	–
–	75.0	1.39	–	1.46	18.0	–	10.6
Segmental motion							
0.0	0.0	14.5	11.3	13.3	–	90.0	99.0
8.83	–	–	–	12.0	–	–	–
1.32	–	9.78	–	–	–	60.9	–
23.08	–	9.20	12.3	–	–	51.1	–
–	26.5	–	–	12.0	–	–	–
–	49.2	11.0	–	–	–	69.2	–
–	75.0	10.1	13.1	11.4	–	63.2	84.6

energy indicate that the reduction of E_a of relaxation processes, in accordance with the stated above, can be explained by the reduction in the density of molecular packing in boundary layers, which contributes to a better realization of mobility both by the side groups and by the segments of polymer chains found in the boundary layers. It might seem strange at first sight that E_a diminishes also for the segmental processes, whereas the position of the corresponding peaks shifts toward higher temperatures. To understand this contradiction, one should bear in mind that the process of relaxation is explained both by the character of interaction in the interface, and by the conformations of molecules in the boundary layer. The information on E_a makes it possible to suggest that the depletion of the conformation number near the interface boundary, the depletion, acting similarly to chain increased stiffness, affects molecular mobility much more than reduction in the density of molecular packing. This concept also finds confirmation in many data obtained in our work which indicate that with the use of fillers that are potentially either able or unable to interact with the polymer, the effects of shifts of the maximum peaks are almost the same, if the comparison is carried out with the correlated thicknesses of the surface layers. It shows that the entropic factor is the determinant in the restriction of mobility of chains in the interface, since with equal thicknesses of the surface layer the shift of the minimums of time of spin-lattice relaxation and of dipole-segmental losses is virtually the same both on the interacting and non-interacting surfaces. Evidently it is of no importance whether the change of conformations is caused only by the presence of the surface or by some degree of bonding of functional groups within the surface (energetic interaction). The latter factor which is essential from the point of view of the strength of the adhesive joints is not significant for the reduction of molecular mobility whose manifestation is not associated with the disruption of bonds in the interface.

It should be noted that all the above examples did not concern intense specific interactions in the interface where the picture will differ from that described. Indeed, it was shown in the work *(73)* that the use of fillers able to interact chemically with the polymer causes a substantially greater change of mobility than in the case of common physical interactions.

From the above point of view it is essential to quantatively divide the contribution to the free energy of activation of the process of relaxation: enthalpy ΔH and entropy ΔS *(74)*.

It was found that ΔH and ΔS of activation depend on the thickness of the surface layer of the polymer (thickness of layer between filler particles). It is clear from the respective equation that the activation entropy in the surface layers increases noticeably as compared with the bulk, while enthalpy decreases insignificantly. It indicates that the entropy factor plays the main role in the change of properties of polymers and in the mobility of chains in the surface layers (Table 4).

Table 4. Enthalpy and entropy of activation of segmental motion

Filler content (a – aerosil, f – fluoro-plaste)	Copolymer		Polyurethane from polyester		Polyurethane from polyether	
	ΔH kcal/mol	ΔS kcal/mol. grad	ΔH	ΔS	ΔH	ΔS
–	13.3	1.10	–	–	–	–
8.83 a	12.0	1.52	–	–	–	–
24.9 a	11.5	1.81	–	–	–	–
26.5 f	12.0	1.50	–	–	–	–
75.0 f	11.4	1.82	–	–	–	–
–	–	–	2.0	0.13	–	–
13.3 a	–	–	1.8	0.20	–	–
13.3 a	–	–	1.8	0.21	–	–
–	–	–	–	–	4.38	1.15
14.8 a	–	–	–	–	3.8	1.79

This increase of activation entropy in the surface layers can be explained by the fact that a greater change in the conformations of chains is required for the transition of molecular segments from one position at the surface to another during the relaxation because of effective greater stiffness of the chain, than for similar movements in the bulk.

It is also appropriate to stress here that, as follows from the respective computations, the principal contribution to E_a is given from the enthalpy component, where as the entropy factor plays the main part in the change of E_a in the boundary layers.

It should be noted that the effects of changes of molecular mobility in boundary layers are characteristic not only for polymers but also for oligomers. It was indicated (*57, 60*) that the decrease of mobility in the boundary layer of a number of oligomers chiefly depends on the decrease of the number of possible conformations, since the effects appeared to be little affected by the nature of the surface. The following considerations can be presented as one more proof of this important assumption (*75*).

The appearance of a loose packing in the boundary layers signifies the transition to a less balanced equilibrium position, to less probable conformation. One should have expected that a lengthy thermal treatment of the boundary layers similar to the effect of such treatment on the approach of glass-like polymers to equilibrium, will also affect molecular mobility. However, the study of PMMA and of copolymer methylmethacrylate and styrene filled with aerosil indicates that the heat treatment leads to a smaller shift of the low-temperature maximums, *i.e.* to a more dense packing, and yet does not affect the position of the high-temperature maximum. The data give evidence to the fact that the increase of the stiffness of chains in the boundary layer reflects the transition of the system to a new equilibrium state, which is characterized by new equilibrium conformations of the chains. It may be suggested that during the heat treatment the chains draw closer to one another without change of conformation, which causes the low temperature processes to shift to the right (distance between molecules diminishes), whereas the conformations of the chains, hence, their mobility, do not change.

In connection with the statements above, let us dwell upon the reports (*76, 77*) in which during the study by the NMR method of the molecular mobility of rubbers filled with carbon black, and during the determination of transition temperatures, no changes or shifts of relaxation processes caused by the filler were detected. In our opinion, as it has already been stated above, when considering the dependence of the shift of Tg on the cohesive energy, that those cases concerned systems with small cohesive energies, due to which the far-reaching influence of the surface could not manifest itself.

In addition, the findings of the work (*73, 78*) indicate that even for elastomers the effect of the filler on relaxation processes may be strong.

Finally, it was shown in (*79*) that the parameters of distribution of relaxation time spectra, calculated from circular diagrams, differ for systems with different filler contents, *i.e.* with different thicknesses of the boundary layers.

Previously we discussed the findings obtained by the method of nuclear magnetic resonance and dielectric relaxation, related to heterogeneous systems with organic fillers.

For the case of polymer fillers it should be expected that the properties of the surface of the polymeric filler will change under the influence of the polymer matrix.

Let us consider the polymer-polymer system in which a stiffer polymer is the filler and the matrix is more flexible. The work (*80*) dealt with the relaxation processes in the boundary layer of an acrylato-epoxy-styrene composition, and of the epoxy resin on the surface of particles of a styrene copolymer with methylmethacrylate. In this case a reduction of segmental mobility may be observed on the polymer surface, since the relaxation process shifts toward higher temperatures. The process of relaxation copolymer segments shifts toward lower temperatures, which implies an increased mobility of the polymeric filler molecules. An analogous picture was noted in the research of those systems by the NMR method. The findings indicate that the more flexible component in the polymer-polymer system stiffens, whereas the stiffer one softens.

We also studied the molecular mobility in the epoxy resin polycaproamide system at various stages of hardening which were simulated by adding various quantities of the hardener. A new phenomenon of the inversion of the mutual interaction of components in such composite materials was revealed. With small degrees of hardening the molecular mobility of resin chains on the surface of the fibrous filler diminishes, and simultaneously a softening of the filler is observed. However, with a highly cross-linked resin, a reduction of segmental mobility of the filler (polycaproamide) is manifested under the action of the greatly hardened compound, as is seen in the T_1 versus temperature curve (Fig. 10). Here the hardened resin itself is characterized

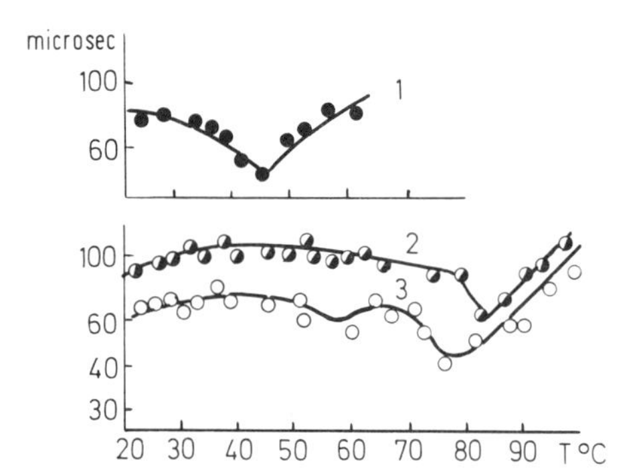

Fig. 10. Temperature dependence of T_1, 1: polycaproamide fiber, 2: epoxy compound, 3: composite (1.1)

by a shift of the relaxation minimum toward lower temperatures through approximately 8°. It means that in filled systems with a polymer as the filler, the latter in some cases limits the mobility of molecules of the binder and simultaneously softens, while in other cases, with a very stiff binder, the latter plays the role of a "solid surface", restricting the mobility of chains of the polymeric filler. In (*81, 82*) the behavior of other polymer-polymer compositions was studied by the method of spin-lattice NMR relaxation with similar results. Thus, for polymers filled with polymeric fillers, in contrast to systems with mineral fillers, both the change of the relaxation properties of the binder and the relaxation properties of the filler should be considered.

From this point of view, of interest is the effect of the introduction of inorganic filler on the molecular mobility of components of a polymeric mixture, components possessing macromolecules of different stiffness. In analyzing a composition of polyvinyl chloride and copolymer styrene and methylmethacrylate, of a copolymer and diacetylcellulose and PMMA and dimethylcellulose filled with aerosil, it was shown (*83*) that as a result of incompatibility of the components and their selective interaction with the surface of the in organic filler, the more polar or more rigid component loses its molecular mobility, whereas the mobility of the less polar component increases in comparison with its mobility in the non-filled mixture. This concept is illustrated by the tabulated data (Table 5). The computation of the parameters of the relaxation time distribution, on the basis of the dielectric relaxation findings, has indicated that in the filled mixture the spectrum of the relaxation time expands, whereas in the mixture without the filler it may be more narrow than in the initial components.

Table 5. Temperature shifts of maximums of dipole-segmental relaxation

System	To higher temperatures in °C	To lower temperatures in °C
1. PVC-copolymer (styrenemethylmetacrylate), shift of PVC maximum	3.5–3.0	–
2. Shift of PVC maximum in the same blend in aerosil presence	8 –9	–
3. Shift of copolymer maximum in the blend	6.65	–
4. Shift of copolymer maximum in the blend in aerosil presence as compared to blend maximum (without aerosil)	–	4–3.5
5. Shift of copolymer maximum in the blend with diacetylcellulose	4.5–5	–
6. Shift of the copolymer maximum in the same blend in aerosil presence as-compared to maximum without filler	–	8–8.5
7. Shift of maximum of copolymer in the same blend in aerosile presence as compared with pure copolymer	–	4–4.5
8. Shift of PMMA maximum in the blend with diacetylcellulose in aerosil presence as compared to PMMA	–	5–6

6. Deformability of Filled Polymer Systems

The modulus of elasticity is most important in specifying the mechanical properties of polymers. In this connection consider briefly the dependence of the modulus of elasticity of a filled polymer on the content of the filler. Most work on those lines was carried out for filled elastomers to which to a certain extent hydrodynamic concepts can be applied, these concepts underlying the inferences about modulus dependence on the filler content.

The principal effect of fillers, poorly interacting with the polymer, on viscoelastic properties is due to the partial filling of the volume by stiff low-mobile inclusions. Particles of the filler are usually big enough, and so the average distance between them, even with a large content of the filler, is great as compared with the root-mean-square distance between the ends of the polymer molecules, for which the Gaussian distribution holds true. Moreover the filler particles are so remote one from another that they cannot be interconnected by one molecular chain. For such systems in which the thickness of the layers between particles is much greater than the interface thickness, the effect of the contribution of interfaces to the total properties of the system may be neglected. With due account given to such premises simplifying the actual picture, many various equations have been obtained which explain the increase of the modulus of elasticity with the introduction of the filler. Smallwood (*84*), proceeding from hydrodynamic considerations, proposed for the evaluation of the modulus of elasticity of a filled system an equation which was analogous to the Einstein equation for viscosity:

$$E^1 = E_{un} \, (1 + 2.5 \, \Phi) \tag{7}$$

in which E^1 = modulus of filled rubber
E_{un} = modulus of unfilled rubber
ϕ = volumetric fraction of the filler.

This equation has proved useful only for small concentrations of the filler, that's why another equation was suggested which depicts the experimental findings more correctly (*85*)

$$E' = E_{un} \, (1 + 2.5 \, \phi + 14.1 \, \phi^2) \tag{8}$$

For non-spherical particles the factor of shape f was introduced as the ratio of the particle length to its crosswise size (*86*), and then

$$E' = E_{un} \, (1 + 0.67 \, f\phi + 1.62 \, f^2 \phi^2) \tag{9}$$

Eilers (*87*) proposed an empirical equation for the same purpose

$$E' = E_{un}\left(1 + \frac{1.25\phi}{1-1.2\phi}\right) \tag{10}$$

Kerner (*88*) theoretically derived the following equation

$$\frac{E'}{E_{un}} = \left\{\frac{G'_F \phi/[(7-5\nu)\, G'_0 + (8-10\nu)\, G'_F] + (1-\phi)/[15\,(1-\nu)]}{G'_0 \phi/[(7-5\nu)\, G'_0 + (8-10\nu)\, G'_F] + (1-\phi)/[15\,(1-\nu)]}\right\} \tag{11}$$

in which E' = composition modulus
ν = Poisson's ratio for polymeric matrix
G'_F = real part of complex shear modulus of the filler

G'_0 = same for polymeric matrix
ϕ_F = volume fraction of filler.

The Kerner equation was modified in the work (*89*)

$$\frac{E'}{E_{un}} = 1 + \frac{AB\phi}{1 - B\phi} \tag{12}$$

in which $$A = \frac{7-5\nu}{8-10\nu} \tag{13}$$

and

$$B = \frac{(E'_2/E_{un}) - 1}{(E'_2/E_{un}) + A} \tag{14}$$

also the filler modulus E'_2 is taken into account. A further modification was performed in the work (*90*) which yielded the expression

$$\frac{E'}{E_{un}} = \frac{1 + AB\phi}{1 + B\chi\phi} \tag{15}$$

in which $\chi\phi$ = function depending on the degree of ultimate filling ϕ_m

$$\chi\phi = \left[1 + \left\{\frac{(1-\phi_m)}{\phi_m^2}\right\}\phi\right]\phi \tag{16}$$

or

$$\chi\phi = 1 - \exp\left(\frac{-\phi}{1-\phi/\phi_m}\right) \tag{17}$$

In recent time the effect of the filler on the modulus of various elastomers has been examined in the work (*91*) with due account to the above-mentioned Kerner's equation, and presuming that owing to the interaction at the interface, part of the polymer is bound by the filler into boundary layers, and as a result, the efficient fraction of the filler should be increased to ϕ_e which is more than ϕ in accordance with the equation

$$\phi_e = \phi(1 + \Delta r/r_0)^3 \tag{18}$$

in which $\Delta r/r_0$ = relative increase of the diameter of particles due to the interaction which can be found from rheological data (see below) or from the comparison of the loss modulus of the filled and unfilled specimens

$$E''/E_{un} = 1 - \phi(1 + \Delta r/r_0)^3 \tag{19}$$

Considering the latter ratio, Kerner's equation can be transformed by introducing $\phi_e = \phi B$ in which B = parameter characterizing the interaction. If $V = 0.5$ and $E_F \gg E_{un}$ the equation will become

$$E'/E'_{un} = (1 + 1.5\,\phi B)/(1 - \phi B) \tag{20}$$

and respectively the mechanical losses will be given by

$$\mathrm{tg}\,\delta_c = \mathrm{tg}\,\delta_0/(1 + 1.5\phi_c B) \tag{21}$$

Evidently, the losses decrease with the introduction of the filler.

Those equations differ from the previous ones in that they consider the actual structure of the filled system and the existence of the boundary layer, though the contribution of the latter to changes of the properties of the polymeric matrix is not estimated here. The equation has been verified on a number of elastomers filled with silica, glass beads, and barium sulphate. The interaction parameter was found from the experimental data by the value of E'', and the Eq. (17). The parameter B, as a rule, decreases with the increase of the filler concentration. However, the introduction of the parameter is of formal character. Its meaning is the relative increase of the size of filler particles owing to the formation of the boundary layer. But then the result is that the thickness of the fixed layer decreases with the decrease of the thickness of the layer between the particles, which is difficult to explain. The dependence of B on ϕ makes it impossible to use the modified Kerner's equation, which does not account for this dependence.

As has been noted, the proposed approach is of interest because it takes into account the existence of the boundary layer, but the dependence of B on ϕ casts doubt on the validity of the approach, particularly if one bears in mind that the thickness of the boundary layer must depend on the impact frequency (*12, 13*). The found magnitudes of B vary within 6 and 1, *i.e.* to a case in which there is no boundary layer. For glass beads in which the diameter of particles is 40 μm on the average, a thickness of the layer of 16–20 μm corresponds to $B = 3$, which by far exceeds the thickness of layers determined by rheological methods.

A theoretical estimation of permanent polymeric systems containing a filler was performed in the work (*92*). Analytical expressions for Lame effective coefficients λ and μ were obtained for medium and large degrees of filling:

$$\mu^* = 2\,(C_1 + C_2)\,\chi(\phi) \tag{22}$$

$$\lambda^* = 2.15\,(C_1 + C_2)\,\chi(\phi) \tag{23}$$

in which $$\chi(\phi) = 0.435\left\{[(\frac{\phi_M}{\phi})^{1/3} - 1] + 1.15\,[(\frac{\phi_M}{\phi})^{1/3} - 1]^2\right\}^{-1} \tag{24}$$

Here ϕ = filler concentration; ϕ_M = maximum degree of packing with the given shape of particles; C_1 and C_2 = Mooney's constants. Respectively, the modulus of elasticity will be given by

$$E = 5\,(C_1 + C_2)\,\chi\phi \tag{25}$$

However, because of a number of assumptions lying in the basis of the calculation, the theoretical values differ by an order of magnitude from the experimental data. A further development of those concepts (*93*) has indicated that the elastic properties of a filled system are determined by the volume fraction of the filler and are little dependent on the size of filler particles.

A generalized equation for the modulus of elasticity for two-phase systems was given by Nielsen (*94*) in the form

$$E'/E_{un} = [1 + (K - 1)\,B\,\phi]/(1 + B\,\chi\phi) \tag{26}$$

in which E' and E_{un} = moduli of the composition and polymeric matrix
B = constant determined by the ratio of moduli of the phases
K = Einstein's coefficient (equal to 2.5 for dispersions of spherical particles in incompressible medium)
χ = parameter determined by the shape of the particles and by the coefficient of fullness (coefficient of filling by the particles of the volume with maximum packing).

The effect of the filler concentration on the dynamic modulus magnitude may be discussed from the point of view of wave propagation in a two-phase medium. In this case (*95*) the theoretical values of the modulus are found from the relationships

$$E' = \{E_{un}\,(7-5\nu) + (8-10\nu)\,E_F - (7-5\nu)\,(E_{un} - E_F)/\phi\}\,/\,/[(7-5\nu)\,E_F + (8-10\nu)\,E_F - (8-10\nu)\,(E_{un} - E_F)\phi] \tag{27}$$

in which E_F = filler modulus
ν = Poisson's coefficient.

An interesting observation has been made in the work (*96*) to the effect that Kerner's equation is not sufficiently reliable for a complex system a glass-like polymer-rubber-like-filler-glass beads. In this case the best results can be obtained if the systems are considered as a single phase of a glass-like polymer with rubber in which glass beads are dispersed. Then the modulus of the medium is first calculated as a system filled with a polymeric filler, following which the calculated value is used for the evaluation of the composite material.

The problem of the effect of the content and size of particles of the filler on the modulus of elasticity is discussed in detail in a number of other works (*97–99*).

No doubt the deviation of the experimental relationships from those described by the equations is associated with the finding that the result of modulus changes cannot be reduced to the hydrodynamic effect of the filler particles only. Thus already in the work (*100*), concerned with the amorphous glass-like state of rubbers, the effect of black on the shear modulus was proposed to be described by the use of the equation

$$G_T = G_1\,\phi_1 + A''\,G_2\,\phi_2 \tag{28}$$

in which G_T, G_1 and G_2 = shear modulus of the system, moduli for the rubber and black respectively, ϕ_1 and ϕ_2 = their volume fractions and A'' = factor characterizing the adhesion of black to rubber.

Payne (*101*) indicates that the dynamic properties of the rubber-black system in a highly elastic state are determined by the following inter-related factors:

1. Structural effect (black structure) specifying the rigidity of filled vulcanizes at small deformations.
2. Hydrodynamic effect of black particles distributed in the viscoelastic medium.
3. Adhesion between black and rubber, the role of the adhesion being increased with the increase of deformation.

On the basis of the above points, Payne proposes the following general equation:

$$G' = G_{un} f(f, \phi) F(A) \tag{29}$$

in which f is the hydrodynamic effect, depending on the factor of the shape of particles and their volume concentration ϕ; $F(A)$ characterized the bonding of the filler and polymer particles. The value of $F(A)$ may be found from the data on the swelling of filled vulcanizates, and reflects the number of effective cross-links as a result of the interaction of polymer molecules with the filler. If $f(f, \phi)$ is found for inert fillers, assuming that $f(f, \phi)$ G'/G_{un} and $F(A)$ is found from the swelling data, a good correlation of experimental data on G'/G_{un} can be obtained calculated with due account $f(f, \phi)$ and $F(A)$.

It follows from the dependencies of the relative modulus (ratio E'/E_{un}) on the content of the filler discussed above, that though E' and E_{un} depend on the temperature, the relative modulus must be almost independent of temperature, despite the prediction by Kerner's theory of its slight rise caused by the increase of Poisson's coefficient with the temperature. According to Nielsen (*90, 102*) the temperature dependence may be associated with the difference in the value of the matrix modulus in a filled system as compared with an unfilled system. It is known that stresses develop around the filler particle in an isotropic medium because of differences in temperature coefficients of expansion of two phases in cooling the material after its formation. Since polymers are noted for a non-linear stress-deformation dependence, the modulus of elasticity decreases with stress. As a result, the polymer has a smaller modulus near the border than the filled polymer, even if the total modulus of the system is higher. The magnitude of stress around the particle decreases with the rise in temperature, and the modulus increases respectively. A corresponding theory has been developed in the work cited which yielded the following equation for the temperature-modulus dependence:

$$\frac{d(E_c/E_0)}{dT} = \kappa\left(\frac{E_{c0}}{E_{10}}\right)\frac{(\alpha_1 - \alpha_2) f(\phi)}{\epsilon_y} \tag{30}$$

in which E_{10} = modulus of unstressed matrix
E_{c_0} = modulus of filled material in the absence of thermal stresses
ϵ_y = elongation in the curve-point stress σ-strain ϵ where $d\sigma/d\epsilon = 0$ and

α_1 and α_2 are the coefficients of thermal expansion of the matrix and filler. The function of $f(\phi)$ may be of various forms, for example

$$f(\phi) = \frac{\text{en}\,\chi^3}{(\chi^3 - 1)} \tag{31}$$

where $\chi = (\phi_m/\phi)^{1/3}$, ϕ = the volume fraction of the filler and ϕ_m = maximum content of the filler in case of dense packing of its particles. The equation proved satisfactory when checked by the filled polyethylene and epoxy resin data. Thus the rigidity of a reinforced material can change greatly with temperature, even if the moduli of the components are little dependent on temperature. The authors have also shown that the temperature dependence of the relative modulus is determined by the ratio of the filler modulus to the matrix modulus which differently affects the relative modulus above and below the glass point.

It becomes evident from this point of view that the equations explaining the dependence of the modulus on the filler content are inapplicable for the temperature transition region. Indeed, it was shown (*103*) that the dependence of the logarithm of the relative shear modulus in the transition region for the epoxy resin-finely dispersed quartz is close to linear, but the slope of the curves is different for different temperatures. This is equivalent to the assumption that one and the same value of the relative modulus at different temperatures corresponds to different contents of the filler ϕ. However, a non-linear dependence on temperature is indicative of a more complex mechanism of modulus changes in the temperature transitional regions.

All the works cited above do not take into consideration the contribution of the change of polymer properties in the interface in relation to the dependence of the modulus of elasticity of the filled polymer on the filler concentration; yet the separation of those effects is essential. This was attempted in the work (*104*) in the analysis of dynamic mechanical properties of polyurethanacrylates, filled with quartz powder, with one and the same volume fraction of the filler and various sizes of its particles. The dependence of E and $t_g\,\delta$ on the filler concentration has been analyzed with particles of sizes that permit the contribution of the surface layers to be neglected. Then, studying the properties of the filled polymer with various sized particles, the effect of the filler associated with its own volume can be excluded, and the effects determined by the surface layers of the polymer may be isolated. In the filler low-concentration region the dependence of the modulus on the filler content is well depicted from the empirical correlation (*103*):

$$E'/E_{un} = 1 + 1.75\phi + (1.75\phi)^2 \tag{32}$$

Figure 11 illustrates the temperature and frequency dependences E and tg δ of the filled polyurethanacrylate. The reduction in the sizes of particles leads to a reduction in the thickness of polymer layers between them, thus increasing the fraction of polymer in the boundary layers. The approximate thickness of the layers calculated on the assumption that the filler particles are of the cubic shape and form a regular body-centerd lattice in the polymeric matrix, with the volumetric concentration of the filler of 0.1, amounted to 3, 5, 7, 40, and 50 microns in the examined fractions.

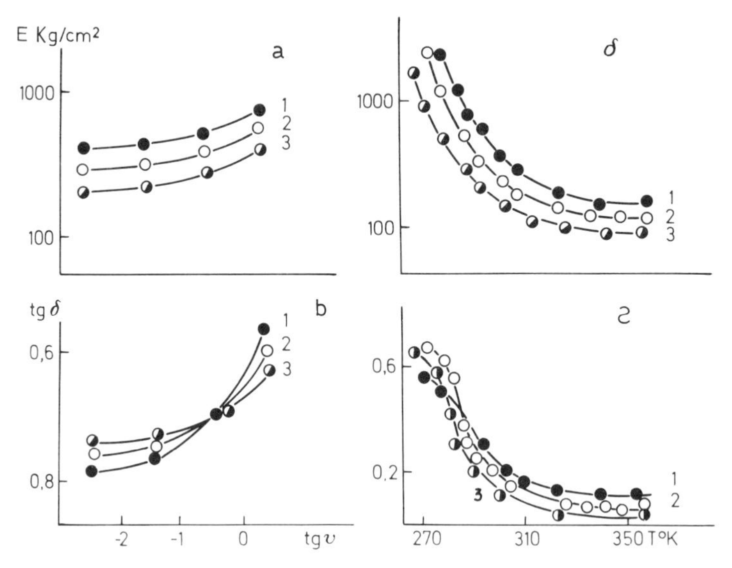

Fig. 11. Frequency (at 293 cHz a, b) and temperature (at 0.004 Hz) dependences (c, d) of E_0 and tg δ (e, f) of filled polyurethanacrylate at $\phi = 0.11$ and various size of filler particles 1–3, 2–7, 3–45 mcm

As is evident from Fig. 11 for specimens with thinner polymer interlayers, the curves are positioned higher and to the right, as compared with specimens with relatively thicker layers. Taking into account the constancy of the concentration of the filler, such arrangement of the curves can be explained by the dependence of the mechanical properties of the polymer interlayers on their thickness. It is obvious that in this case E_{un} stands for the elasticity modulus of the binder whose properties have been changed under the action of the filler surface. Using the formula, the equilibrium modulus $E_{0\infty}$ of the polymer interlayers can be calculated from the experimental equilibrium values of the high elasticity modulus of the filled specimens.

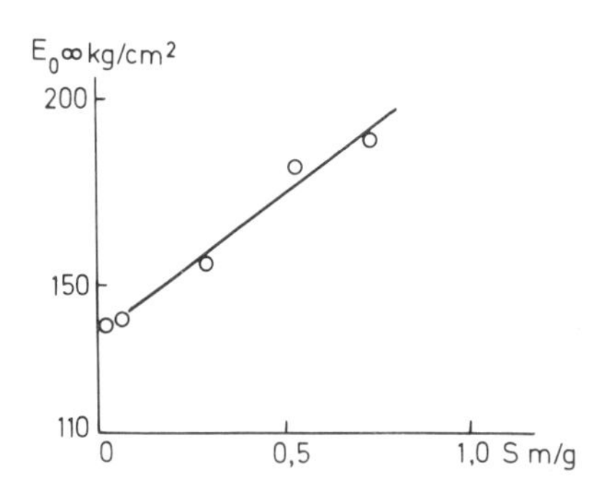

Fig. 12. Dependence of E_∞ of filled PUA on specific surface area of the filler ($\phi = 0.11$)

Figure 12 illustrates the dependence of the equilibrium high-elasticity modulus $E_{0\infty}$ of polymer interlayers on the filler surface calculated from the formula. One can see that the increase of the specific surface of the filler brings about an increase of $E_0\infty$. Since this effect is more expressed in specimens with a higher specific surface fillers, a logical conclusion follows that it is associated with the presence of the boundary layer of the polymer on the filler surface which (layer) either does not take part in the deformation or possesses properties differing from the properties of the polymer remote from the filler.

Taking into consideration that the thickness of the polymer interlayers between filler particles by far exceeds the length of the macromolecule and square mean root distance between its ends, for a filled polymer the Gaussian distribution of distances between the ends of macromolecules must hold true. Temperature dependence of relaxation time in such system can be described with the aid of Williams-Landell-Ferry (WLF) equation whose applicability to filled systems was checked by many scientists. Therefore, for a further analysis of the experimental findings, use has been made of the WLF equation and of the free-volume conception.

The shift factor α_T was determined from the temperature dependences E and tg δ using the formula proposed by Ferry (*105*):

$$\alpha_T = \frac{\eta T_s \delta_s}{\eta_s T \delta} \tag{33}$$

in which η = internal friction coefficient
T = temperature in °K
δ = specimen density

The index "S" is given to characteristics of the specimen at reduced temperature T_s 50° above the position of the losses maximum tg δ.

It was shown that the shift factor α_T in the polymer with a coarse filler is virtually the same for both large and small concentrations of the filler, whereas for a fine-grained filler the curve is steeper and lies lower. It indicates that the boundary layers take part in the deformation process, and that their properties differ from those of more remote layers.

Using the values of α_T and the formula from the work (*105*)

$$\lg \alpha_T = B\left(\frac{1}{f} - \frac{1}{f_g}\right) \tag{34}$$

we have calculated the free volume fractions in the filled specimens at various temperatures. For the calculations, use has been made of the universal value of the fractional free volume at the glass transition temperature f_g = 0.025. It was found that the temperature-dependence curve of the free volume fraction for specimens with a more developed surface was higher than for specimens with a coarse-grain filler. It points to the increased fractional free volume in the boundary layers which agrees well with the previously established reduction of packing density in boundary layers (*1*).

To obtain a more unambiguous interpretation of the findings of the mechanical examinations we have carried out a calorimetric study of the same objects. Temperature dependences of the specific heat C_p in the transitional region were obtained, and on their basis the fraction φ_n of the polymer found in the boundary layer was analyzed.

Figure 13 illustrates the fraction of polymer in the boundary layer versus the specific surface of the filler with a constant concentration of the latter. As is evident from the drawing, the boundary layer fraction rises as the filler surface increases. Comparing this fact with the increase of modulus of elasticity with the increase of

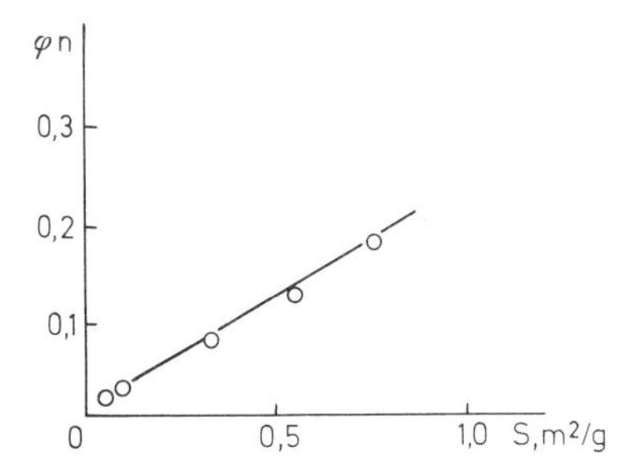

Fig. 13. Dependence of surface layer fraction on filler surface at $\phi = 0.11$

the filler specific surface, a conclusion can be drawn that changes in properties of the filled polymer are due to the transition of part of the polymer to the boundary layers with accompanying changes of properties.

To estimate the effect of the filler on the polymeric medium, a concept has been introduced in the work (*106*) about a critical concentration of the filler at which the entire mass of the polymer changes to glass under its influence. This concentration ϕ_{cr} is found as the point of intercept of the log modulus to filler concentration curves in the regions above and below the glass point.

The introduction of the critical concentration concept makes it possible to obtain curves for filled compositions which are invariable with respect to the filler nature. For plotting such curves the concentration of the filler is replaced with the reduced value which defines the degree of change of the composition at the given concentration ϕ and varies from 0 to 1. This reduction helps to obtain generalized dependences of a number of properties (including modulus, glass transition point, etc.) on the reduced concentration, irrespective of the nature of the filler. At ϕ_{red} the polymer changes into the surface layer state, and then the thickness of the layer can be readily found from ϕ_{red} as $\delta = V/S$ (in which V = polymer volume, S = total surface of the filler at $\phi/\phi_{cr} = 1$). The values thus obtained for different fillers are within a resonable range (80–400 A). However, since ϕ_{cr} is defined from the data on the mechanical properties, the thicknesses reflect just those characteristics.

The work (*107*) also states the need to estimate the characteristics of filled systems from the equations taking into account the characteristics of the interface whose properties substantially differ from those of the polymer hardened without the filler. Indeed, it follows from the above-stated that the change in the structure of the surface layer, as compared with the properties in the bulk, is a factor determining the physico-chemical properties of filled polymers. However, up till now there have been no direct data which would make it possible to estimate the contribution of the property of the surface layer in a filled system to the properties of the composition material as a whole. It is necessary to calculate a filled system as a three-element one, consisting of the filler, interface with changed properties, and of an invariable binder. In this area of interest is an attempt made in the studies (*108, 109*). The authors estimated the critical content of the filler ϕ_{cr}, above which no changes in the density of the polymeric fraction of the composition occur, and suggested that under those conditions all the polymeric component is in the interphase layer. Such a system may be represented as a two-component one. For the system under study (thermosetting binder-inorganic filler) the authors have found that the density of the polymeric composition is higher than the additive value which was attributed to the interphase layer of a higher density. The density of the interphase layer was found from the function of density of the filler fraction at the ϕ value, and this value was used to estimate the volume fraction of the interphase layer from the additivity equation for a three-dimensional system.

The latter can be used for the calculation on the mechanical properties of a system, if at ϕ_{cr} the binder modulus is replaced with the interphase layer modulus. Then the composition modulus will be defined from the equation:

$$E_c = \frac{E_f E_b E_l}{E_f E_e \phi_b + E_b E_e \phi_f + E_f E_b \phi_l}$$

in which E_c, E_f, E_l, E_b are elasticity moduli of the composition, filler, interphase layer, and binder; ϕ_f is the filler fraction ϕ_b = binder fraction, and ϕ_l = interphase layer fraction.

Thus it has been established that the interphase layer characteristics differ greatly from those of the binder, and these differences are associated with the nature of the polymer and filler *i.e.* with adsorption interaction, hardening conditions, etc.

7. Viscoelastic Properties of Polymer Compositions with Dispersed Polymeric Fillers

Mechanical properties of polymeric compositions were studied by many scientists (*e.g.*, see 110–*113*). That is why this section will deal only with some general problems mainly related to relaxation properties.

Dynamic mechanical properties of polymers filled with polymeric and non-polymeric fillers can be described for the case of absence of interaction between the components on the basis of mechanical models proposed by Takayanagi (*114*). The composition can be schematically illustrated by his sketch 14a, in which the system type is at the left and the equivalent model at the right. The upper drawing refers to a homogeneously distributed dispersed phase, and the lower drawing, to a heterogeneously distributed phase. If the phase a is dispersed in the phase ω two equivalent models for systems I and II are possible, respectively (Fig. 14b). The complex modulus of elasticity for those models will be given as

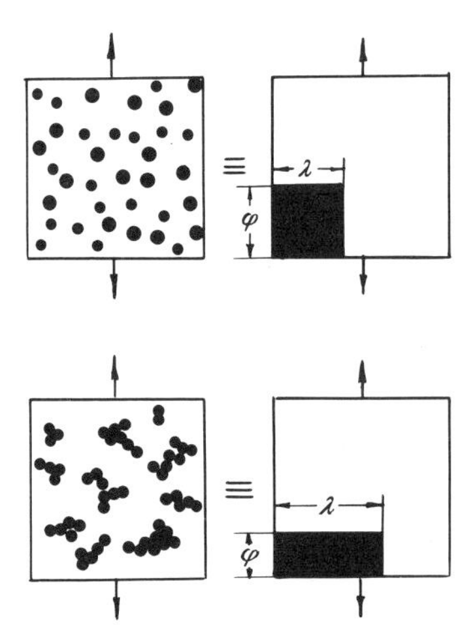

Fig. 14. a) Scheme of two-phase mixture and of equivalent model [Ref. (*114*)]

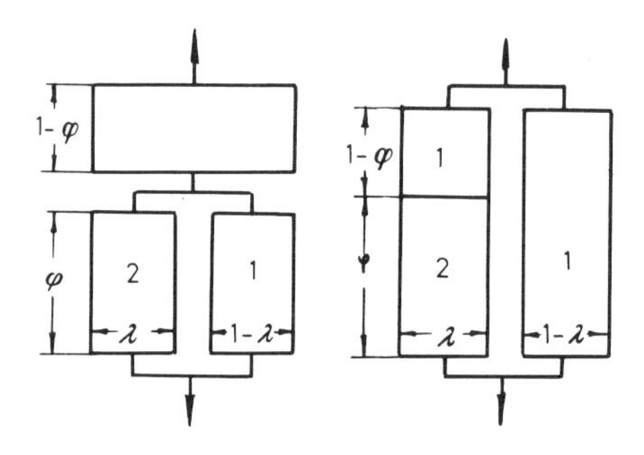

Fig. 14. b) Mechanical models for calculation of modulus from equivalent model [Ref. (*114*)]

$$E_{\mathrm{I}}^* = \left[\frac{\varphi}{\lambda E_{\mathrm{a}}^* + (1-\lambda) E_\omega^*} + \frac{1-\varphi}{E_\omega^*}\right]^{-1} \tag{35}$$

$$E_{\mathrm{II}}^* = \lambda\left[\frac{\varphi}{E_{\mathrm{a}}^*} + \frac{(1-\varphi)}{E_\omega^*}\right]^{-1} + (1-\lambda) E_\omega^* \tag{36}$$

in which E_{a}^* and E_ω^* are the complex moduli of the two phases, magnitudes λ and φ correspond to the fraction occupied by phase a in the thickness and length of the specimen (see diagram) and $\lambda\varphi$ is equivalent to the volume fraction of phase a. Values λ and φ may be called the composition parameters, since they vary for different types of composition, $\lambda\varphi$ remaining constant. The greater φ, the closer the model to the common paralled connection of the elements; and the greater λ, the nearer it is to the series connection of the elements.

It has been established that those equations can be applied to many systems in which two components form two separate non-interacting phases. Thus, in (*115*) the model of Takayanagi was employed to denote the dynamic modulus and mechanical losses of butadiene-styrene rubber reinforced with polystyrene particles 400 Å in size. It has been established that in the high-elastic region the filler sharply increases the modulus, but has little effect on it in the glass-like state region. No changes in the glass-transition points of the components have been revealed in the mixture, as compared with the pure components. The behavior of the system is adequately described with the aid of the equations based on the model, without depicting the effect of the fillers on segmental mobility of macromolecules of the rubber phase. The same model was advantageously employed for the polymer-glass sphere system (*167*).

Proceeding from the assumption that the indeterminancy in the disposition of the elements in the series paralled connection makes operations with the model difficult, another model was proposed (*116*) which removes a number of difficulties without the introduction of additional parameters. The model is schematically represented in Fig. 15. Its geometric parameters a and b are unequivocally connected with the volume fraction of the dispersed phase, as the other component, by means of the ratio

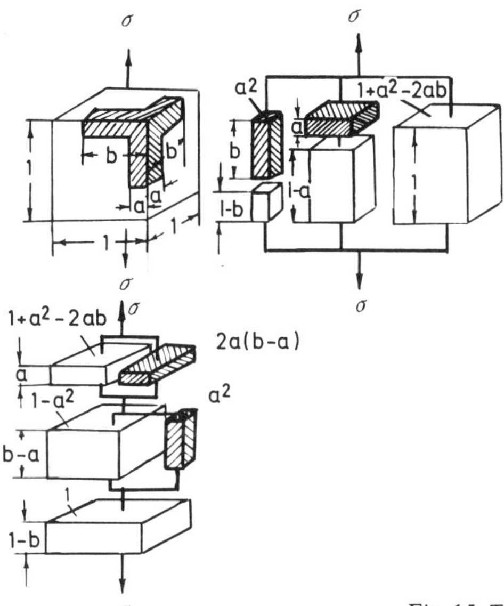

Fig. 15. The "Isotropic" models A and B [Ref. (*116*)]

$$V_2 = a^2\,(3b - 2a) \tag{37}$$

Figure 15 illustrates two methods of arrangement of the elements of the model, which are the limiting cases of the series-parallel (A) and parallel-series (B) connections. Evidently the isotropy of this model is more preferable from the physical point of view, as compared with Takayanagi's model. The authors stress that the proposed models represent only a phenomenological method of describing experimental data, and that in connecting the model, empirical rules are used, which is justifiable only when correlating the results of theoretical estimations with the experiment.

The moduli in various models are found from the following equations:
Isotropic model A

$$E = \frac{a_2}{(1-b)/E_1 + b/E_2} + [(1-a)^2 + 2a(1-b)/E_1 + 2a(b-a)/](1-a)/E_1 + a/E_2 \quad (38)$$

Isotropic model B

$$E = \left\{ \frac{a}{a(2b-a)E_2 + (1-a)^2 E_1} + \frac{(b-a)}{a^2 E_2 + (1-a)^2 E_1} + \frac{(1-b)}{E_1} \right\}^{-1} \quad (39)$$

E_1 and E_2 are complex moduli. Their components elasticity and loss moduli are found for the mixture by a direct substitution of the respective values in those relationships with a subsequent division into real and imaginary parts. In the cited work both the models were correlated with the experimental data and with calculations by the equation of Takayanagi, and the conclusion was made that the formula (38) agrees with the experimental data best.

The theory of viscoelastic properties of geterogeneous composites of the polymer-polymer type was developed on the basis of the well-know equation of Kerner (*117*). This equation is represented for the shear modulus G in this form

$$\frac{G}{G_m} = \frac{(1-V)G_m + (\alpha + V)G_i}{(1 + \alpha V)G_m + \alpha(1-V)G_i} \quad (40)$$

where ν is the volume concentration of the second phase, m and i are the matrix and filler indexes and $\alpha = 2(4-5\nu_m)(7-5\nu_m)$ in which ν_m is the matrix Poisson's ratio. A similar expression can be derived for the complex dynamic modulus

$$\frac{E^*}{E^*_m} = j\,\frac{(1-V)E^*_m + \beta(\alpha + V)E^*_i}{(1 + \alpha V)E^*_m + \alpha\beta(1-V)E^*_i} \quad (41)$$

where $\beta = (1 + \nu_m)/(1 + V_i)$,
$j = (1 + \nu)/(1 + \nu_m)$
in which ν is Poisson's ratio for the composition.

A composite is described with the aid of the same Takayanagi's models which were discussed above. The equivalence of both the models enables them to be described with the aid of one equation. If viscoelastic Poisson's ratio ($\nu^* = \nu^{i} + i\nu''$) is equal to $\nu^* = \nu^1 = \nu$, then the dynamic shear modulus

$$G^* = \left[\frac{\varphi_1}{\lambda_1 G^*_i + (1-\lambda_1)G^*_m} - \frac{(1-\varphi_1)}{G^*_m} \right]^{-1} \quad (42)$$

of the 2-model equations for the moduli may be reduced to the Eq. (40) if we find the parameters of the model $\varphi_1 = V(1 + \alpha)/(\alpha + V)$ and $\lambda_1 = (\alpha + V)/(1 + \alpha)$ after an experimental check up of the dependence of the modulus on the volume fraction

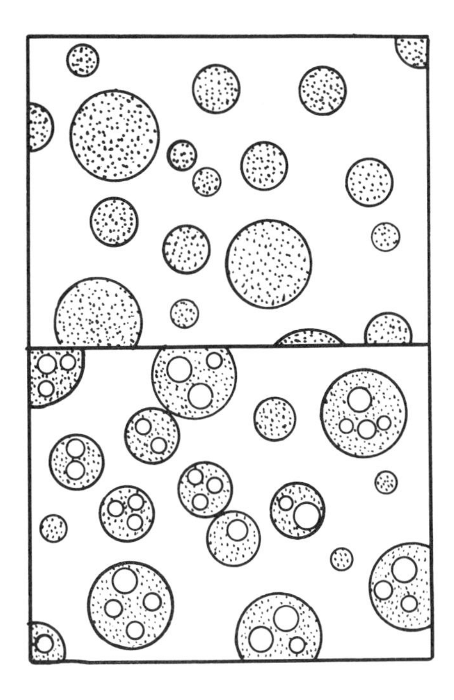

Fig. 16. Scheme of the composite material with simple (a) and complicated (v) inclusion [Ref. (*117*)]

of the second component, of Poisson's ratio for the matrix and of other properties of the phases. Two types of inclusions have been considered: a simple inclusion of one polymer in another (Fig. 16), and complex inclusion in which part of the polymer, forming the dispersed phase, occludes part of the continuous phase. Properties of such particles can be calculated from the equations, and then the values obtained again substituted into the same equations for the estimation of the composite properties. A good relationship was established between the experimental and calculated values for systems which are particles of a soft filler in a rigid polymeric matrix. The estimations indicate that part of the complex modulus associated with the included component depends on the volume concentration of the dispersed phase, and the part associated with the matrix depends both on its concentration and morphology. The considerations were checked against specially synthesized model systems (*118*). During the calculations with the aim of correctly depicting the results, the values of ν_m taken varied from 0.2 to 0.5. For the model systems (dispersion acrylate latex in PMMA and the like) whose structure was estimated by the electronic microscopy technique, calculations of the dependence of the modulus of elasticity on the composition were carried out using the Kerner's equation. It has been found that in a number of cases the effective fraction of the volume dispersed phase should be introduced into the theoretical equations instead of the real volume, taking into account the dependence of the former on the temperature. The authors emphasize the important role of the phase inversion which may lead to a change in the process of mechanical loss dependence on the temperature. The model concepts were taken as the basis for the study of the effect of morphology on the properties of polymeric compositions in the work (*119*) in which morphology is understood to be the character of distribution of filler particles and their sizes in the polymer-matrix place.

A drawback of all the considered model concepts is that the possible interaction between the components in the inferface is neglected.

As a consequence of the above statements, changes in the relaxation properties of the polymers themselves, properties resulting from the adsorption interaction should be taken into account, as well as the possibility of the formation of additional cross-links in the physical structural network, as a result of the polymer-surface interaction (*1*). The possibility of the formation of structures as a result of the interaction of the filler particles with each other should also be taken into account. Thus viscoelastic properties will be determined not only by the existence of two independent phases, as considered in the model discussed above, but also by many other factors. Of much interest from this point of view is the work (*120*) in which a three-phase model accounting for the formation of the layer with properties, differing from those in the bulk, is proposed for the case of the effect of a filler with a high surface energy on the formation of transcrystalline regions of polyethylene on its surface.

The research of viscoelastic properties of filled polymeric systems is very important for the estimation of the behavior of articles made from them under operating conditions and under impact of considerable loads.

Now consider general regularities in the viscoelastic behavior of filled polymers against their chemical structure and flexibility of chains.

Here the theory of viscoelastic properties stated in a number of manuals will not be discussed.

The principal regularities of viscoelastic properties of polymers are revealed in analysis of their temperature-frequency dependence. Such properties are described in the theory of reduced variables (*105*). Williams, Landell, and Ferry worked out a method of transformation of temperature and frequency scales with the aid of which experimental data, specifically the dynamic modulus, can be put on one generalized curve, covering a very wide range of frequencies and temperatures (WLF-method). In a number of studies carried out up till now the applicability of the WLF equation to filled systems, mainly to rubbers, has been proved (*121–126*).

Temperature dependence of relaxation and retardation time of various filled vulcanizates can also be depicted with the aid of the WLF equation

$$\lg \alpha_T = \frac{-8.86\,(T-T_s)}{101{,}6 + T-T_s} \tag{43}$$

in which α_T = reduction coefficient which defines the shift of the given characteristic in the frequency log dependence curve

T = temperature in the experiment

T_s = reduction temperature which is an empirically selected magnitude.

The application of the WLF theory to filled rubbers has indicated that the filler has little effect on the temperature-frequency dependence, and that T_s is approximately linear-dependent on the filler content (*127*).

In most studies the WLF method was used to plot generalized curves of viscoelastic functions for specimens with varying content of the filler. Validity of principle of temperature-frequency superposition for filled polymers has been demonstrated. It has been established during the experiments that in the comparison of specimens with different degrees of filling at the respective temperatures (*i.e.* at reduction temperature T_s for the unfilled polymer and $T_s + \Delta T_s$ for each of the filled specimens), the retardation and relaxation spectra agreed well at short time periods and filler volume fractions up to 0.2, while with greater filling degrees the relaxation-time spectrum maximum increases, whereas the average relaxation-time monotonously decreases.

Calculations of the relaxation-time spectra for black-filled rubbers were also performed in the study (*128*) with similar results. It was demonstrated that with increasing activity of the black, the planar is the relaxation-time spectrum, *i.e.* the more significant is the role of the relaxation time of filled systems, which is connected with the restriction of mobility of macromolecules in the strengthened structures of the polymer.

Generally, the effect of the filler on the relaxation-time temperature dependence can be described as the increase of the values T_s (*124, 126, 127, 129*).

The actual mechanism of relaxation processes in filled compositions is very intricate and insufficiently studied. It is conjectured that the active filler causes the increase in the number of relaxation elements with greater relaxation periods. Bartenjev and Vishnitskaya (*130*) have ascertained the existence of three relaxation elements connected with the chains relaxation proper with the detachment of rubber chains from the filler particles and with regrouping of the filler particles (filler relaxation). In the work (*131*) the nature of high-temperature peaks of mechanical losses of black-filled rubber was studied and the conclusion was made that their existence is connected with the inversive destruction of black structures and of the polymer structural lattice associated with them. Lyalina, Zelenev and Bartenev (*132*) have also studied the effect of the black surface nature on the relaxation processes and have confirmed the above-stated concepts. They also observed the multistage relaxation mechanism in black-filled rubbers and indicated that the process activation energy associated with the relaxation in black-rubber complexes is dependent on the filler content (*133*). The effect of deformation of the structure formed by the filler was accounted for in the work (*134*). It was demonstrated that the principle of the temperature-time superposition is inapplicable with high filler contents when additional relaxation mechanism come to light. The appearance of a compacted structure of the polymer (*135, 136*) caused by its interaction with the surface of the filler particles making up the structural network, leads to the appearance of a number of new features in the relaxation behavior, features associated with a change in molecular mobility. The analysis of the relaxation spectrum for the black-filled vulcanizate of the butadien-styrene polymer (*137*) has shown that the introduction of the black filler increases the height of the spectra, and that the character of the spectra is dependent on the black energy characteristics. In the case of small deformations the spectrum of the vulcanizate with black, characterized by a homogenous surface, shifts toward greater relaxation time, whereas for an active black with a heterogeneous surface the spectrum falls abruptly in the large relaxation-time region. In the case of great deformations ($> 50\%$) the spectrum of vulcanizates with active blacks shifts toward the large relaxation time region the more, the greater the reinforcing action of the black. Under such conditions the increase in the height of the relaxation spectrum and its shift toward greater relaxation periods is associated with the compactness of the structure and presence of strong polymer-filler bonds. The increase in temperature accelerates the relaxation processes and brings about a destruction of weak bonds, which cause the spectrum to diminish in height.

A molecular theory of the relaxation properties of filled elastomers has been developed by Sato (*138*) on the basis of the statistical concept of rubber-like elasticity. He has derived expressions for the estimation of Young's modulus stresses and mechanical losses of filled polymers.

While the question of the viscoelastic properties of filled rubbers has been examined in greater detail, the dynamic properties of thermoplastics and thermosetting plastics have not been sufficiently studied.

Relaxation processes in filled cross-linked phenolic resin have been investigated in the work (*139*). The authors stress that changes in the interaction between the resin and filler against changes in the temperature or speed of the deformation lead to changes of the system's mechanical properties with the filler introduction. On the basis of the findings they calculated the relaxation-time spectra and estimated activation energy of relaxation processes. There was a good agreement between the theoretically calculated and experimental spectra of relaxation time, though they did not reveal a simple correlation of the results obtained with different rates and temperatures in the experiment. Unfortunately the authors did not examine the physical essence of the processes which occur in the interaction with the filler, and confined themselves to a formal application of the viscoelastic theory to the filled system, and to the establishment of certain differences in the relaxation-time spectra.

The effect of the filler on the dynamic mechanical properties of PVC compositions filled with aerosil and black has also been analyzed (*140, 141*). As has been established, the existence of strong cohesion in the boundary layers changes the mobility of the chains and the spectrum of relaxation properties. The filler brings about an increase of G^1 and slightly affects the character of its temperature dependence. The reduction in temperature increases the ratio of the moduli of filled and unfilled polymers in the glass point region. The changes in the spectrum are explained by the formation of boundary layers with an increased free volume. The authors also stress the importance of the filler effect on the properties of the polymeric matrix. The effect of the filler size on the dynamic properties of epoxy resin compositions with aluminium powder is discussed as well (*142*) and a conclusion is made that the increase in the dynamic characteristics is indicated with the optimum size of the particles.

It should be noted that the formation of the filler structural lattice in thermoplastics can also affect the viscoelastic properties (*143*). The viscoelastic properties of and PMMA filled with aerosil and glass fiber have been examined in (*149*) in which the role of the filler and polymer nature was demonstrated as well as their interaction in response to those properties. The dynamic properties of filled polyethylene, polypropylene, and of a number of other polymers have been investigated in (*72*) where they are treated from the point of view of changes in the molecular mobility. A number of other reports on those problems have been published (*145–148*). The principal findings of these studies confirm our concepts related to effects of molecular mobility on the viscoelastic properties. However, in contrast to the studies of filled resins, the filled thermoplastics were examined only by the use of common methods of dynamic analyses without any attempts at a detailed examination of the relaxation spectra and quantitative description of the results with the aid of the WLF theory. Only one report (*149*) is concerned with the examination of the spectra of relaxation time for amorphous PS and PMMA, filler presence. The change in aerosil. These spectra are also noted for a shift toward greater periods and an increase of the contribution of medium relaxation periods in triller presence. The change in the spectrum shape is approximately proportional to the filler surface area.

From the point of view of the effect of the formation of interphase transition layers on the viscoelastic properties, of interest is the report by Romanov and Ziegel (*150*) who studied the dynamic mechanical properties of filled elastomers, using the ethylene-vinylacetate copolymer with polyamide, PS, and PMMA as the fillers taken in various contents by volume. They have attempted, on the basis of the temperature dependence of the moduli of the components and the modulus of the composition, to calculate the parameter which specifies the interaction of the components by the agency of the increase of the volume of the particles owing to the adhesion of the adjoining interphase layer. It has been found that this parameter is constant above the glass point of the polymer matrix and reduces at temperatures below this point, and that it is independent of the filler content. The dynamic mechanical properties of a number of incompatible polymeric compositions have been investigated in a number of works by Zelenev *et al.* (*151, 152*). It has been shown that the main region of relaxation of the components expands as a result of the diffusion distribution of the components. However, typical for all the cases is the manifestation of two relaxation regions whose positions are constant, though the heights of peaks of mechanical losses may change with a change in the mixture composition. Zelenev also observed in the mixtures butadiene-styrene rubber with PS a shift of the peak toward lower temperatures. Additional relaxation peaks have been displayed which are associated with the "defrosting" of the mobility of the segments in the transitional layer zone (for the elastomer-PVC systems).

From the point of view of a general description of properties of systems filled with polymeric fillers, of much interest is the report (*153*). In contrast to the common-type filled systems, in which the filler is introduced into the bulk of the polymer matrix, a system is under study in which the immobilization of the polymer regarded as the filler, was performed by impregnating the surface layer of cellulose specimens with diluted solutions of the polymeric filler. Used in the experiment were incompatible systems, which make it possible to determine the properties of the "bonded" surface polymer, the properties which reflect the adhesive interaction. Copolymers of styrene and acrylonitrile with butadiene have been analyzed with the determination of the mechanical properties of the starting materials and of the compositional material. Changes in the rubber-glass transition points have been determined on the basis of the data on temperature dependence of the imaginary part of the complex elasticity modulus with different amounts of the polymer introduced into the surface layer. It turned out that the temperatures (T_g) are always higher for the polymer in the surface layer than for the same polymer within the bulk. The increase in the glass transition point is regarded as a measure of the interphase interaction. The interaction depends on the material composition and on the composition of the polymer and increases as the parameters of solubility of the polymer and cellulose draw together. The experimental findings were compared with mechanical characteristics computed on the basis of mechanical models for polymeric mixtures in which the components do not interact.

An interesting method of producing systems filled with a polymeric filler is the method of polymerization of a monomer in which a polymer is dissolved incompatible with the resulting polymer. The relaxation properties of such a system will depend on the initial concentration of the polymer in the monomer. In the work (*154*) a number of such systems based on PBMA-PMMA have been examined by determining the parameters of the Kolrausch relaxation equation

$$E(t) = E_{\infty} + E_0 \exp E\left(\frac{t}{\tau}\right)\kappa \tag{44}$$

in which $E(t)$ = modulus at a time point t
E_{∞} = equilibrium modulus of elasticity
E_0 = relaxating part of the modulus
τ = relaxation time.

Systems have been studied with various contents of PBMA and sizes of its particles and various distributions of particle size. The difference in sizes of particles was caused by a diverse viscosity of the reaction medium with the dissolved polymer which affected the character of aggregation of the PBMA molecules.

A sharp reduction in the equilibrium modulus of elasticity corresponds to the introduction of a low molecular PBMA. The minimum of the equilibrium modulus and the maximum of the relaxation time correspond to the maximum sizes of the particles. However, the non-linear change of the module is explained by the authors by the finding that the relaxation properties are affected not only by the properties of the two components but also by the processes occurring at the interface. The increase in the size of the particles brings about a reduction of the interface- area values. For this reason the effects associated with the change of molecular mobility at the interface will be manifested in those areas in which the particles are smaller and their number greater, *i.e.* either with small or great concentrations of PBMA in the system. A delay in the change of the equilibrium modulus occurs in the reagion of 30 to 40% of PBMA in which an abrupt increase in the number of particles and in the surface area takes place. The plateau region corresponds to the maximum of size and minimum of number of the particles and is due to the presence in this region of the concentration of a great number of small particles, owing to which a solution of PBMA in PMMA actually makes up a continuous phase in this region. The authors conclude that the character of distribution of the components in the system has a significant effect on their relaxation behavior which is the function of both the composition and distribution of particles of the dispersed phase in the polymer matrix.

Heterogeneous polymeric systems on the basis of two polymers can be also realized in the form of interpenetrating polymeric networks (*155, 156*). Such systems are interesting in that they consist of two continuous, not chemically inter-connecting polymeric networks which cannot be separated because of mutual penetration of segments of one network into the cells of the other. The appearance in IPN of a transitional region caused by the incompatibility of the components increases the heterogeneity of such systems and can be noted on their viscoelastic properties (*157*).

Changes in the concentration of the second network has a non-monotonous effect on the apearance and position of the temperature curves of the real part of the complex modulus of shift of the mechanical losses tangent. The effects are associated both with the change in the relation of phases and fraction of the transitional region, and with the effects of the first network on the properties of the second network, the effect which is similar to the effect of the filler on the properties of polymers (*1*). It has also been found that the introduction of the second network has a great effect on the relaxation spectrum, the effect which is similar to the effect of the filler (see below) and is associated with the reduction of molecular mobility in the second network because of the adsorption interaction with the first network.

Interesting aspects of the effect of fillers on the relaxation behavior were noted in (*158*) illustrated by experiments with segmented polyurethanes. In this case the presence of segregations of blocks of different chemical structure, typical for block-copolymers, causes the appearance in such systems of selective interactions with blocks of different nature, and brings about a complex effect on the temperature curves of the viscoelastic characteristics, as well as on the modulus-filler concentration curve.

The findings of the research of viscoelastic properties discussed above are based on the study of properties of the entire filled system. It is assumed that with a sufficiently high concentration of the filler, having a large specific surface, the thickness of the polymer layers between particles becomes commesurate with the thickness of the polymer boundary layers on the surface of the filler, and the boundary layers contribute appreciably to the general mechanical behavior of the filled polymer (§ 1). However, quantitative data can be obtained with the aid of an equation depicting the dependence of viscoelastic functions of a filled polymer on the concentration and viscoelastic characteristics of the components in the absence of a specific interaction between them. Though there are such correlations, no one of them is sufficiently correct. This may be because of a substantional heterogeneity of the strain-stress state of the polymeric layers because of their irregular thickness and differences in the structure at different distances from the interlace. Moreover, some of the layers may be under tensile strain, while others may be under shear or compression. There is a possibility that some part of the polymer owing to virtual non-deformability of the mineral fillers will be under conditions of all-round compression or tension.

Homogeneity of the polymer strained state can be maintained by the use of the method of examining the mechanical properties of a pile of layers (*159*). The concept of the method, as is known, consists of the measurement of the total strain of n-number of the boundary layers of the polymer applied to thin high-molecular plates stacked in a pile. Knowing the total deformation of the pile and the deformation of the high-molecular plates, the deformation of the polymeric layers can be readily calculated. Since in the study of linear viscoelastic properties the strains measured on a layer of, say, 10 μm thick will be very small, a special technique of measurement of dynamic mechanical properties of a pile of layers has been elaborated (*159*).

The study and comparison of the viscoelastic properties of multi-layer polymer specimens with various thicknesses of the polymer layer will make it possible to elucidate the effect of the nature of the solid surface on the layer thickness and properties of the surface layer. Let us discuss some of the findings. The viscoelastic characteristic of solidified epoxy resin with the thickness of layers of 7.5 to 100 were examined by the use of quartz plates.

It follows from the data on the temperature and frequency dependence of the real part of the complex shear modulus E' will rise and tg δ drop with the reduction of the layer thickness. This phenomenon can be explained as follows: The modulus changes are associated with stiffening of the polymer in the thin layer near the solid surface. Suppose that part of the layer in the immediate vicinity to the surface differs from the more remote layers and serves as a "rigid" filler for those layers. In such a case the higher values of E' can be attributed to the increased concentration of the filler, or to a reduced concentration of the polymer in the interlayers owing to the transition of polymer particles into the boundary layer. For such a model the thickness of the boundary layers can be estimated from the experimental data by the effect of the polymer concentration on the modulus. The analysis of the calculated thickness of layers has indicated that they are temperature dependent. This dependence can be explained on the assumption that the temperature dependence of the boundary layer of the binder is close to that for a pure binder, but is shifted toward the higher temperature region. As a result, the more spectacular differences in properties of the boundary layer and the rest of the polymer should occur in the transition-state temperature region, when most of the polymer has already passed into the state with low values of the modulus, whereas the boundary layer is still in the high moduli state. It actually means that not the thickness of the boundary layer but its contribution to the mechanical behavior as a layer with changed mechanical characteristics is temperature dependent. Indeed, in the glass-like state the modulus of the binder is large and the efficient contribution of the boundary layer with a high modulus is not very significant. All this similarly holds true in the highly elastic state, and the difference is manifested only in the transition state with a high temperature dependence of the properties. For this reason the transition state is most convenient for the study of properties of the boundary layer and their differences from bulk properties. Experiments prove that the efficient layer thickness temperature-dependence curve passes through a maximum whose position corresponds to the mechanical loss maximum. Assuming the thickness of the boundary layer equal to the maximum efficient thickness and using the formula for composite specimens with successive work of the layers (*160*) we have

$$G'_e = \left(\frac{\varphi_s}{G'_s} + \frac{\varphi_0}{G_0}\right)^{-1} ; \operatorname{tg} \delta = \left(\frac{\varphi_s}{G'_s} + \frac{\varphi_0}{G'_0}\right) \left(\frac{\varphi_s}{\operatorname{tg} \delta_s G'_s} + \frac{\varphi_0}{\operatorname{tg} \delta_0 G'_0}\right)^{-1} \tag{45}$$

in which φ_s = boundary layer concentration with shear modulus G'_s in the polymer layer

φ_0 = concentration of the bulk portion of the polymer with modulus G'_0 and φ'_e = efficient characteristics of the polymer layers calculated from the ratio

$$\frac{G'}{G'_e} = \frac{1}{(\varphi_s + \varphi_0)} \tag{46}$$

The temperature dependences G_s' and tg δ_s have been evaluated. It was shown that in the transition-state temperature region the shear modulus of the boundary layers is higher and the maximum of losses is shifted toward higher temperatures.

The findings indicate that the polymer properties in the layer are perhaps not uniform. To throw light on this non-uniformity in (*161*) temperature dependences of mechanical losses of the same epoxy resin polymer have been studied in its successive proximities to the support by the use a technique based on the determination of the temperature dependence of mechanical losses of a three-layer cantilever rod in which the internal layer is a metallic foil and the two external layers are the polymer under examination (*162*). The initial thickness of the layer in the subsequent experiments was reduced either mechanically (polishing) or by way of sublimation under the action of low-temperature plasma (*163*). It has been found that for very thin layers of the polymer, the maximum is shifted toward higher temperatures. Figure 17 illustrates the dependence of the temperature of the maximum on the polymer layer thickness. The findings indicate the non-uniformity of the structure of layers in agreement with the findings of the experiments (*2, 5*).

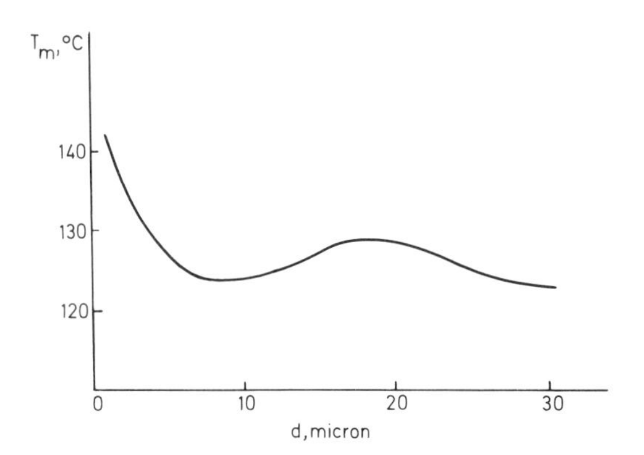

Fig. 17. Dependence of temperature of maximum of mechanical losses on the thickness of polymer layer

In this way the data given above show that the viscoelastic properties of the systems under study are much dependent on the existence of the macro- and microheterogeneity of structure and on the interphase interaction in the interface. The latter, by virtue of the reasons explained in Section 1, changes molecular mobility of polymer chains, and consequently, the entire complex of relaxation and mechanical properties.

8. Effect of Inorganic and Polymeric Fillers on Relaxation Spectra

As has already been indicated, the restriction of molecular mobility because of adsorption interaction leads to an expansion of the relaxation spectrum. It was shown in (*55, 58*) on the basis of the data on the distribution parameter calculated from dielectric measurements by the method of circular diagram of Cole-Cole (*62*).

The increased average time of relaxation on the surface of solid particles (*1, 46, 47*) and the increase of the relaxation spectrum are connected with changes in the surface layer structure (*2*). From this point of view a systematic reduction in molecular mobility and the increase of average relaxation time with the increase of the interface must occur until all relaxants with greater relaxation periods are excluded from the relaxation process as a result of a strong binding of large numbers of polymer molecules with the surface, as takes place during the vulcanization of elastomers. Therefore, it can be assumed that regular changes in the character of mechanical relaxation time spectra will occur with changes in the concentration of the filler. Since this problem has not been specially investigated, we have carried out the study of viscoelastic properties of filled polymers under dynamic loading conditions (*164*) with the aim of checking the hypothesis. We have studied frequency and temperature dependences of the complex shear modulus G^* and mechanical loss tangent tg δ for an epoxy resin composition with different concentrations of the filler. Relaxation-time spectra have been computed from the resultant dependences of the real part of the complex shear modulus on the strain frequency by the method of Ninomia-Ferry with the aid of punch cards.

The curves of the spectral function of H for relaxation periods more than 10 sec. and for various concentrations of the filler are illustrated in Fig. 18. It is clear that

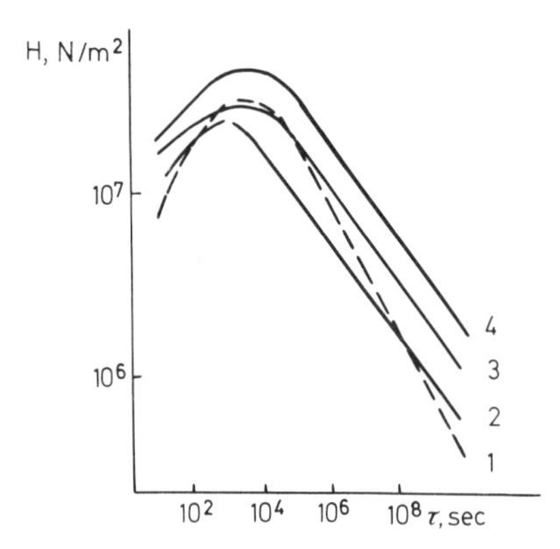

Fig. 18. Relaxation spectra of filled with quartz powder polyepoxy resin at 80°, 1: initial, 2: ϕ = 0.04, 3: ϕ = 0.31, 4: ϕ= 0.44

with small concentrations of the filler (φ = 0.04 by volume) the relaxation-time spectrum undergoes perceptable changes only in the short-time region, somewhat shifting toward the small relaxation-time side.

An increase in the concentration of the quartz filler (curves 2 and 3) also results in an expansion of the spectrum and its shift toward larger periods. In addition, the spectral function *H* increases with the increase of the filler concentration, and the slope of the spectrum linear portion in the filled specimens is less prominent than in the unfilled ones.

Taking into account that the quartz relaxation-time spectrum does not overlap the binder spectrum, the indicated changes in the spectra are evidently associated with changes in the properties of the polymer matrix, changes stipulated by the effect of the filler surface and by changes in the conditions of its strain caused by the influence of the bulk of the high-modulus filler.

Let us dwell upon the changes of polymer deformation conditions in the presence of the filler. It should be noted in first place that a high-modulus filler is practically not deformed during the deformation of the filled specimen, since its modulus is thousands of times greater than that of the binder. That is why the deformation of the filled specimen occurs only on account of deformation of the binder. The presence of a sufficiently large number of non-deformed filler particles results in the relative strain of the polymer layers between the filler particles being able to exceed many times the total deformation of the filled specimen (*165*). Owing to this, stresses in polymer layers also have a large amplitude. In accordance with (*166*) an increase in stresses leads to a reduction of the relaxation periods, which can explain the initial shift of the spectrum toward smaller periods. To confirm this supposition we performed an experiment in which the amplitude of deformation of the specimen was increased several times.

Relaxation-time spectra versus various deformation amplitudes are Illustrated in Fig. 19. It is seen from a comparison of the positions of spectral curves that, indeed, the increase of the deformation amplitude substantially shifts the spectrum toward

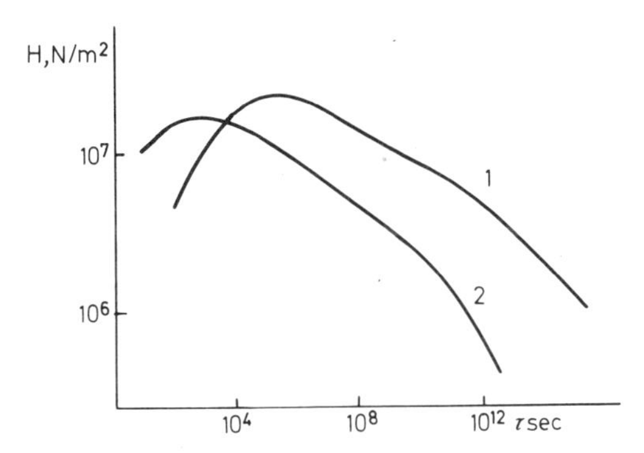

Fig. 19. Relaxation spectra at various deformation amplitudes at 80°, 1: ϵ = 0.01, 2: ϵ = 0.05

shorter periods. It signifies that the presence of the quartz filler would have shifted the spectral curves to the left and the shift would have increased as the filler concentration increased. However, it is evident from Fig. 18 that such a shift does not occur. Moreover, the spectra shift to the right, as the filler concentration increases, expand, and the slope of the linear part of the spectral curves changes. This gives reason to suggest that the increase of the filler concentration enhances the role played by the filler surface which reduces binder mobility, thus shifting the spectra toward greater periods. This surface effect not only compensates for the shift of the curves to the left, expected to be due to the filler incompressibility, but even shifts the spectral curves to the right. Particularly noticeable are the shifts in the larger relaxation-time region. It means that the filler surface has the greater effect on the limitation of mobility of the longest structural elements, which brings about an increase in the average relaxation time. A slight expansion of the spectra toward smaller relaxation periods may be interpreted as a result of an increased imperfection of the polymer network due to the presence of the filler as well as due to a loosening of the molecular packing in the polymer boundary layer (*1, 2*).

Thus, the experiments with quartz filler have indicated that the shift of the spectral function curves for polymers filled with a high-modulus filler is due to the action of at least two causes, one of which is a change of the deformation conditions in the presence of filler particles (shifts the spectrum to the left) and the other is the effect of the filler surface on the properties and structure of the polymer matrix (shifts the spectrum to the right).

It seemed expedient to continue with the experiments, omitting the effect of one of the factors. It turned out to be convenient to exclude the changes in the polymer matrix deformation conditions by choosing the filler approaching in its mechanical properties those of the binder. Most suitable in this respect was the powder of the same solidified epoxy resin, which was used as the binder.

Figure 20 illustrates relaxation-time spectra for specimens with such filler versus various concentrations of the filler. For the sake of comparison the same Fig. 20

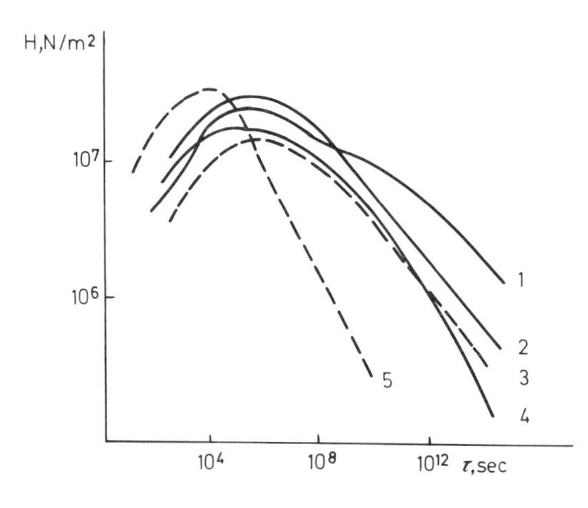

Fig. 20. Relaxation spectra of epoxy resin filled with epoxy filler at 80°,1: $\phi = 0.05$; 2: $\phi = 0.10$, 3: $\phi = 0.40$, 4: $\phi = 0.25$, 5: relaxation spectra of the filler

represents a curve for a specimen from which the filler was made (epoxy resin solidified in the absence of the filler). In the analysis of the experimental results, attention is drawn to the significant shift of the spectral curves toward greater relaxation time as compared with the spectrum of epoxy resin solidified without the filler. An expansion of the spectral curves and a change in their slope also occur in the presence of the filler. It is significant that the shift and expansion of the spectra in this case is appreciably greater than for the specimens with quartz filler. This in associated with the exception of the factor of non-deformability of the filler, which causes the effect of the filler surface on changes in the properties of the boundary layers of the binder hardened on this surface to be manifested more clearly.

From the point of view of the increase in the concentration of the boundary layers with the increase of the filler concentration one would have expected a regular arrangement of the spectral curves. Indeed this regularity can be traced in Fig. 20; however, it turned out to be more complex than could have been expected. Really, it is seen directly in Fig. 20 that with the increase of the concentration of the filler the right-hand parts of the spectrum shift toward greater periods. It indicates that the spectra of the boundary layers of the binder hardened in the presence of the filler are substantially not identical to the spectra of a polymeric filler of the same nature. With the aim of simplifying the analysis, consider the concentrational dependence of the average relaxation time of the systems which have been studied.

Figure 21 illustrates a filler concentration versus average relaxation-time curve. As is seen in Fig. 21, the dependence is of a non-monotonous nature with the mini-

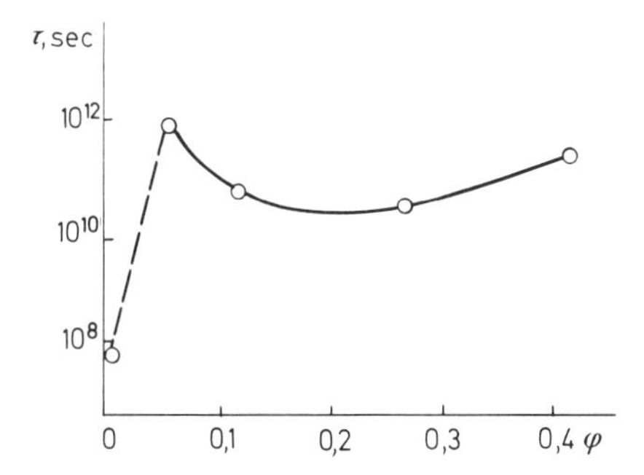

Fig. 21. Dependence of the average relaxation time on polymeric filler concentration at 80°

mum in the region of 15 to 20% of the filler concentration. Along with the reasons stated at the beginning of the article, this nature may be due to the fact that the increase in the filler concentration reduces the average relaxation time of the system, inasmuch as the average relaxation time of the filler itself is less than that of the binder solidified in the presence of the filler. At the same time, the increase in the filler concentration increases the part of the boundary layers, and the increase in the concentration of boundary layers with greater relaxation time increases the average relaxation time of the system. As a result of the counteraction of those two factors, the average relaxation time of the polymer filled with a polymer filler becomes a a non-monotonous function of the filler concentration. The noted non-monotonousness of the change in the viscoelastic properties can also be associated with the non-monotonousness in the changes of the structure of polymer boundary layers on the surface of particles occurring with changes in their volumetric part and thickness.

The study indicates that the filler surface indeed exerts an appreciable and complex influence on the spectra of relaxation time of filled polymer systems, owing to changes in the conditions of deformations of polymer binder layers and changes in the properties of the polymer boundary layer.

9. Principle of Temperature-Time-Filler Concentration Superposition and Its Application to Polymer Compositions

Since the restriction of molecular mobility in boundary layers is equivalent to an increase in the chain rigidity or to a formation of an additional number of bonds in the polymer structural network, it may be assumed that the introduction of the filler in its effect is similar to a reduction in temperature or to an increase in the frequency of deformation. It follows that together with the well-known principle of temperature-frequency superposition, a principle of frequency-temperature-concentration superposition should also be maintained. This principle may be formulated in the following way: An increase in the concentration of the filler in a system leads to an increase in the real part of the complex elastic modulus, which is similar to an increase in the frequency of deformation or a reduction in the temperature of the latter. Prior to our studies experimental data on this question were stated in one work only (*167*).

The effect of the filler concentration was studied (up to 84% – in copolymer of butadience with acrylic acid) in the presence of 0–84% solid filler on the temperature dependence of parameters of mechanical dynamic properties. For the first time a principle was stated in that work to the effect that the filler influence may also be described by the reduction method similar to temperature reduction. For this, use was made of an equation equivalent in its form to the WLF equation but with different numerical values of the constants C_1 and C_2. Its most general result is the stress-strain representation in a twice-normalized state-by temperature and by concentration of the filler. However, not all the parameters can be represented in the concentration-invariable form by the use of the same method of filler concentration reduction. Specifically, the effect of the filler at the initial section of the stress-strain curve is greater than was predicted by the general reduction method. The relative strain at rapture can also be represented in the concentration invariant form, if the vertical shift of the experimental curves is used in addition to the common horizontal one. The author has proposed empirical formulas describing the concentrational dependence of the reduction coefficient.

In the reports (*103, 104*) we have substantuated in detail the concentration-temperature and concentration-time analogies for the description of properties of polymers filled with dispersed fillers, and proved the feasibility of plotting generalized $\log G'$ versus $\log \omega \alpha_T$ curves for specimens containing various quantities of the filler. The dependence $\log \alpha_T = f(T-T_g)$ furnishes some proof of the applicability of the Williams-Landell-Ferry method. It was of importance to investigate the effect of the filler concentration on the average relaxation time of the polymer matrix in the filled material. With this aim, a generalized tangent of angle of mechanical losses to deformation frequency curve was drawn. It was shown that the maximum of mechanical losses shifts toward lower frequencies with the increase of the filler concentration. Making use of the relationship $\text{tg}\, \delta = f(\omega)$ and bearing in mind that the relaxation time $\tau = 1/\omega_m$ (in which ω_m is the frequency corresponding to the maximum of mechanical losses) the functions of $\log \tau = f(\phi)$ were evaluated. The relation is close to a linear one which indicates an exponential dependence of the relaxation time upon the filler concentration. This makes it possible to draw a conclusion about the existence in filled polymers of the filler concentration-time superposition. Indeed, a characteristic shape and position of the $\log G' = F(\log \omega)$ curves with various ϕ make it possible to believe that the WLF method is applicable to those systems. A generalized curve of $\log G' = f(\log \omega\alpha_\phi)$ can be plotted after reducing the term to the lowest concentration of the filler and introducing the concentration-reducing coefficient (α_ϕ). The dependence curve of log $\alpha_\phi = f(\phi)$ also confirms the applicability of the WLF method. Obviously, the reduction coefficient α_ϕ represents a ratio of average relaxation periods in specimens with various contents of the filler

$$\alpha_\phi = \frac{\tau_{\phi 1}}{\tau_{\phi 2}} \tag{47}$$

α_ϕ also characterizes the shift of the curve of log $G' = f(\omega)$ on the complex modulus similarly to the increase in the deformation frequency. The existence of such superposition makes it possible not only to examine the effect of the filler on dynamic and mechanical properties, but also to significantly expand the frequency range for predicting the properties. Since average relaxation periods increase with the increase of the filler concentration, an important problem arises about the effect of the filler on the temperature dependence of the relaxation periods. Fig. 22 illustrates the temperature dependence of the real part of the dynamic modulus with various concentrations of the filler. Apparently, an increase in the filler content shifts the curves of log $G' = f(T)$ toward higher temperatures. Those data can be used for deriving the concentrational dependence of the modulus at various temperatures (Fig. 23). As is seen, the curves of log $G' = f(\phi)$ have a characteristic shape and are

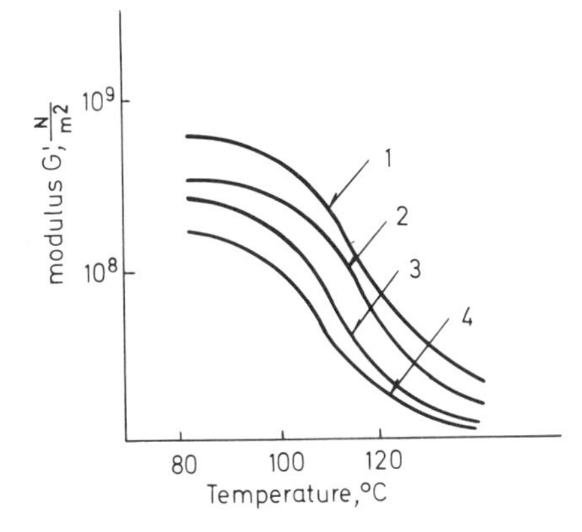

Fig. 22. Temperature dependence of real part of complex modulus at various filler content, 1: 0.44, 2: 0.32, 3: 0.11, 4: 0.045

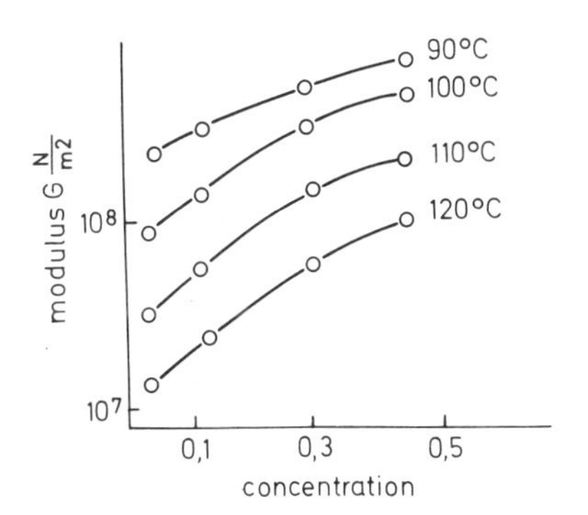

Fig. 23. Concentration dependence of real part of complex modulus in the α-region

equidistant, which makes the WLF method also applicable to them. The generalized curve is represented in Fig. 24. The log $\alpha_{T,\phi} = f(T)$ dependence is nearly linear. Thus,

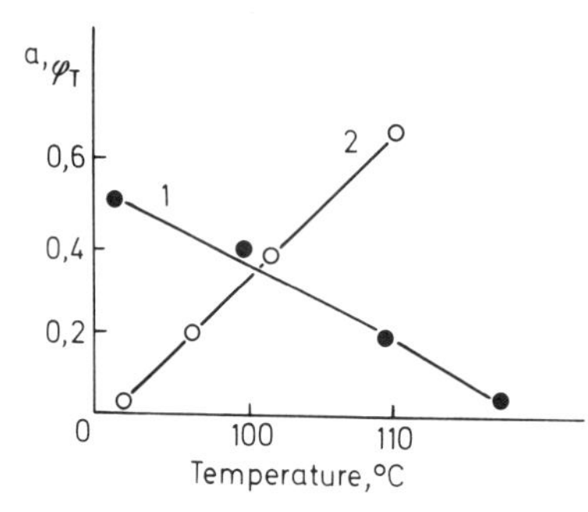

Fig. 24. Temperature dependence of shift factor $\alpha_{T,\phi}$
1 - for composition with quartz filler; 2 - for polystyrene filler

in the latter case as well, an increase in the filler concentration results in a rise of the real part of the modulus equivalent to a reduction in temperature which can be characterized as the concentration-temperature superposition. Its existence makes it possible to expand the feasibility of predicting the effect of the the filler on the properties.

The possibility of the frequency (temperature)-filler concentration superpositions results from the exponential dependence of viscoelastic properties on the filler concentration; their physical significance is that, owing to a considerably higher modulus of the mineral filler, the latter undergoes virtually no deformation, and this changes the deformation conditions of the polymer matrix. In filled specimens the amplitude of deformation of the matrix increases substantially with the rise in the filler content, the total deformation of the specimen remaining the same, which can also be a cause of the increase of the stress and modulus (*168, 169*). As was already noted, the transition of part of the polymer into the surface-layer state with a change of the mechanical characteristics and reduced molecular mobility is another important cause of the modulus increase.

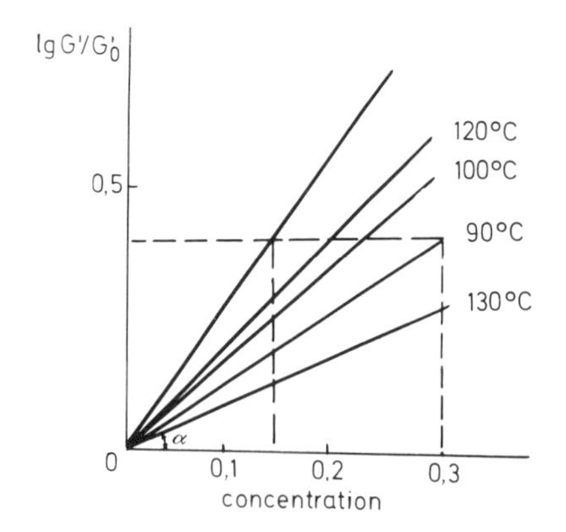

Fig. 25. Concentration dependence of relative modulus

The existence of such a rigid or undeforming layer is in fact equivalent to an increase in the apparent size of the particles or volumetric concentration of the filler. Consider the magnitude of the relative modulus which is the ratio of the modulus of the filled specimen G', to unfilled G'_{un}. G'_{un} is a value obtained by the extrapolation of G', to the zero concentration of the filler. Some of the data are given in Fig. 25. They indicate that the dependence of log (G'/G'_{un}) on the filler concentration is linear, but the slope of the curves is different for different temperatures. Hence, one and the same values of the relative modulus at different temperatures correspond to different concentrations of the filler. For example, the value of log $(G'/G'_{un}) = 0.4$ at 110 °C it corresponds to $\phi = 0.12$, at the same time at 90 °C $\phi = 0.3$. If we attribute this phenomenon to different contributions of the surface layer to the total properties at different temperatures, then changes in the thickness of the surface layer with temperature can be approximately evaluated, Indeed, with a constant number of filler particles their volumetric fraction in the polymer is proportional to the volume of the particles $V = d^3$ (d = diameter of particle with the surface layer). Then for spherical particles

$$\frac{\phi_1}{\phi_2} = \frac{V_1}{V_2} = \frac{d_1^3}{d_2^3} \tag{48}$$

in which the indices 1 and 2 apply to two temperatures. If we denote the slope of the curve of log (G'/G'_{un}) V_s ϕ as α, then

$$\text{tg}\,\alpha = \log\,(G'/G_{un})\,\phi \tag{49}$$

whence

$$d_1^3/d_2^3 = \varphi_1/\varphi_2 = \text{tg}\,\alpha_2/\text{tg}\,\alpha_1 \tag{50}$$

Hence

$$d_1/d_2 = \sqrt[3]{\text{tg}\,\alpha_2/\text{tg}\,\alpha_1} \tag{51}$$

Taking as a basis for comparison the size of particles with the surface layer at 90 °C we calculated the ratios d_1/d_2 for different temperatures. The dependence is characterized by the maximum in the region of the maximum of mechanical losses. This coincidence is evidently associated with the fact that at this temperature the time of the experiment is comparable with the average time of relaxation of the polymer matrix (as was already indicated above, the thickness of the surface layer is dependent on the frequency of action).

It is apparent that at a temperature corresponding to the maximum of mechanical losses, the relaxation periods in the surface layer exceed the time of the experiment, and therefore that layer cannot be deformed substantially. At the same time at greater distances from the interface, the relaxation periods are comparable with the experimental time of action, and for this reason the total deformation is dependent on the deformation of those more remote layers.

Of interest is the application of the principle of temperature-time superimposition or the method of reduction of variables also to systems containing a polymeric filler and polymer mixtures. The applicability of this method to the desdription of heterogeneous polymer mixtures was demonstrated in (*170*), which made it possible to enlarge the frequency range in which the characteristics of the material could be found. However, it was established that the coefficient of reduction α_T for two-phase systems can be the function of time, and that superposition is observed only within a limited temperature interval. The transition from one reduction temperature to another brings about a change in the shape of the curves (*171*).

In considering the relaxation properties of polymers filled with mineral fillers we ascertained the existence in the filled polymers of the filler frequency-concentration superposition. In (*164*) the existence of such superposition was checked for systems containing polymeric fillers. The advantage of such an approach is that it is based on general theoretical concepts, and is not associated with the choice of any definite model of structure of the composite material and with the need to specially account for the interaction on the interface and for the existence of transitional layers.

Dynamic and mechanical properties of the epoxy resin powdered polystyrene (hardening with polyethyleneamine) have been examined. The average size of the filler particles was 50 μ. The filler concentration varied from 5 to 50% by volume. The specimens were deformed according to the harmonic law within the temperature range of 80 and 140 °C. In the experiments the amplitudes of deformation and stress were changed and so was the phase angle between them, and the data were used to evaluate temperature dependences of the real part of the complex shift of shear modulus G' of the filled specimens. Figure 26 illustrates the frequency dependence

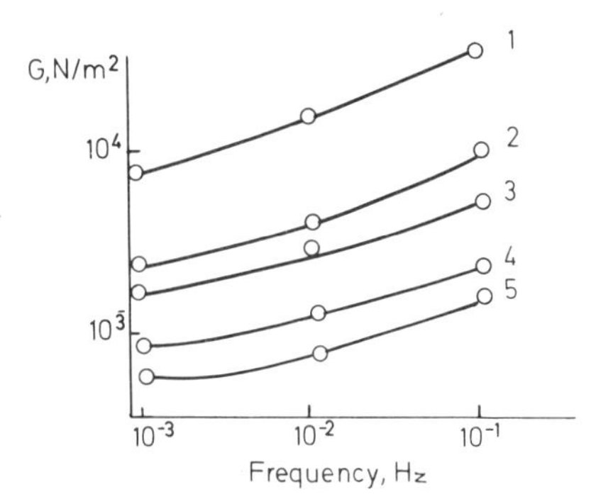

Fig. 26. Frequency dependence of the real part of complex modulus at various filler concentration. 1: epoxy resin, 2: epoxy resin + 5% PS, 3: epoxy resin + 10% PS, 4: epoxy resin + 25% PS, 5: epoxy resin + 50% PS

of the real part of the complex modulus for the specimens with various concentrations of the filler. It is evident directly from Fig. 26 that the increase in the concentration of the polymer filler shifts the curve of $\log G' = f(\omega)$ toward higher frequencies. Similar curves have been obtained at other temperatures in the region of the α-transition of the epoxy resin.

Generalized curves reduced to the filler concentration of $\log G' = f(\log \omega\, \alpha_\phi)$ were plotted in accordance with the data of Fig. 26 and the concentration shift factor $\log \alpha_\phi$ was then found. The generalized curves of $\log G' = f(\log \omega \alpha_\phi)$ for temperatures in the transition region are represented in Fig. 27. The fact that the generalized curves are smooth and the experimental points fit well into the curves (matching of shape of adjacent curves) indicates that the criteria of applicability of the method hold true and that the frequency-concentrational superposition does take place.

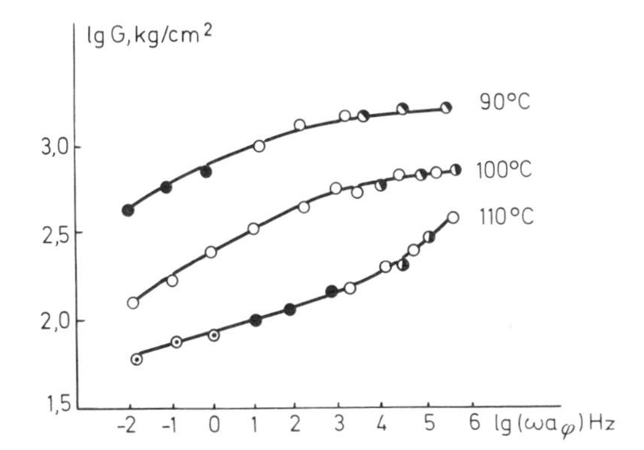

Fig. 27. Generalized curves $\log G' = f(\log \omega \alpha_\phi)$ for transition region: 1: 110°, 2: 100°, 3: 90°

The filler concentration dependence close to the linear one is one more reasoning in favor of this conclusion. It is evident from the method of obtaining the $\log \alpha_\phi$ value that it characterizes the shift of the curves along the frequency axis $\log \omega$ with various concentrations of the filler, *i.e.* the difference between the frequencies at which one and the same value of the real part of the complex shift modulus is observed on specimens with different concentrations of the filler. It follows from the type of dependence of $\log \alpha_\phi = f(\phi)$ that an increase in the concentration of the filler results in the same reduction of the real part of the complex shift modulus as a reduction in the logarithm of deformation frequency. However, in contrast to a system with a rigid filler, the superposition here is of an opposite character, *i.e.* an increase in the filler concentration is equivalent to a reduction in the deformation frequency and not to its increase (frequency derivative is negative) (Fig. 24).

Several feasible causes of such superposition in the "polymer-polymeric filler" system can be named.

The most probable cause is an excessive imperfection of the binder polymer network, the binder which polymerizes in the presence of the polymeric filler. This is manifested by a reduction in the equilibrium modulus of the rubber-like elasticity of the epoxy resin binder with an increase of the filler concentration, despite a complete exhaustion of the epoxy groups in the specimens, which is confirmed by the use of infra-red spectroscopy.

An increase in the network imperfection also takes place in the case with the quartz filler, but because of a high surface energy of the filler there is a great probability of the formation of adhesive bonds of macromolecules with the filler surface which is equivalent to an increase in the network density.

Another cause, in our opinion, is associated with the temperature dependence of the mechanical properties of polystyrene, which is characterized by excessive softening in the region of transistion-state temperatures of the epoxy matrix. Naturally, the increase in the filler concentration in this case must also reduce the real part of the complex modulus of elasticity of the system. The reduction of the shear modulus with an increase in the polystyrene concentration and a reduction in the average relaxation time can be interpreted as a result of an increase in the segmental mobility in the epoxy resin matrix. Therefore, the temperature dependence of the free volume for specimens with different contents of the filler was calculated by the use of the formula (34) in which $B = 1$ and $f_g = 0.025$, after determining the temperature dependence of the shift factor α_T from the experimental curves. It was found that the free volume fraction of the "polymer-polymeric filler" system increases with the increase of the filler concentration. Assuming that the free volume of the system is the additive function of the free volume of the components and using the temperature dependence of the polystyrene fractional free volume, we calculated the temperature dependence of the epoxy resin matrix free volume fraction. The calculations were performed by the use of the formula

$$f_{syst} = f_{pst}\phi_{pst} + f_e\phi_e \tag{52}$$

With an increase in the filler concentration the fractional free volume of the epoxy resin becomes higher than in pure epoxy resin. In this way, the increase in the segmental mobility in the system can be explained perhaps by a more loose structure of the polymer in the epoxy resin matrix. Since the increase in the free volume fraction occurs with an increase in the concentration of the boundary layers, it may be assumed that the increase in the free volume fraction occurs chiefly in the polymer boundary layer.

Since the epoxy resin binder is a rigid three-dimensional polymer, the findings should be interpreted from the point of view of a change in the density of the network cross-links. With this aim we carried out measurements of the equilibrium modulus of rubber-like elasticity of filled specimens. It was shown that the network density with the polymeric filler concentrations reduces.

Temperature dependences tg δ in the region of glass-transition temperature of the polymer matrix were obtained for the same specimens filled with PS. The increase in the concentration of the polymeric filler shifts the maximum of losses toward lower temperatures. It indicates that in specimens with a greater degree of filling the molecular interaction weakens as a consequence of a looser packing of segments in the boundary layer.

10. Conclusion

The results discussed above indicate that in heterogeneous polymeric materials, in accordance with general physico-chemical principles stated in section 1, the viscoelastic properties of a polymeric matrix are strongly dependent on the type of heterogeneity and on the nature of the interphase interactions. In all cases the changes in molecular mobility and structure of the polymer in the boundary layer close to the interface are the principal causes of changes in the viscoelastic characteristics. In those cases when the changes affect most of the polymeric matrix bulk, the boundary layer properties will contribute most to the viscoelastic properties. In their turn those properties will be determined by the nature of the interphase and intermolecular interactions, flexibility of the macromolecules and, for cross-linked polymers, by the degree of cross-linking. Resulting from these factors, changes in the density of cohesion energy and free volume as well as changes in the conformational set of macromolecules associated with energetic and entropic factors lead to changes in the manifestation of molecular mobility of segments and side groups, as well as molecular aggregates.

Thus the viscoelastic properties of the matrix undergo alterations as a result of:

1. Changes in conformations of chains near the interface, the effect of which is transferred through intermolecular interaction to more remote layers.
2. Formation of additional adsorption bonds of macromolecules and of submolecular structure with the surface.
3. Changes in the structure of the boundary layer caused by formation of the polymeric matrix in the presence of solid of the non-polymeric or polymeric nature.

The first two factors manifest themselves in the existence of superposition "temperature-frequency-filler concentration". The latter factor depicting the dynamical mechanical properties leads to non-linear effects of changes in viscoelastic properties.

All the stated factors must be taken into account in selecting the optimum conditions for operation and processing of polymers and in predicting the properties of heterogeneous compositions.

Acknowledgement. I thank Dr. V. Babich and Dr. L. Sergeeva for valuable help in preparation of materials and Dr. T. Todosyichuk for help in manuscript preparation.

11. References

1. Lipatov, Yu.: Physical chemistry of filled polymers, (Russ.) Kiev 1967.
2. Lipatov, Yu., Sergeeva, L.: Adsorption of polymers. New York: J. Wiley 1974.
3. Lipatov, Yu., Fabulyak, F.: Dokl. Akad. Nauk SSSR **205**, 685 (1972).
4. Kuleznev, V., in: Mnogokomponentnye polymernye sistemy. Moscow: Chimiya 1974.
5. Lipatov, Yu.: Pure a. Appl. Chem. **43**, 273 (1975).
6. Kozlov, P., Korostylev, V.: J. Phys. Chem. (Russ.) **31**, 653 (1957).
7. Rabinowitch, A.: Vysokomolek. soed. **1**, 1062 (1959).
8. Frisch, H., Madfai, S.: J. Amer. Chem. Soc. **86**, 3561 (1958).
9. Oosawa, F., Asamura, S.: J. Chem. Phys. **22**, 1255 (1954).
10. Lipatov, Yu., Babich, V.: Vysokomolek. soed. **10B**, 848 (1968).
11. Babich, V., Lipatov, Yu.: Mechanika polymerov **3**, 548 (1969).
12. Zarev, P., Lipatov, Yu.: Vysokomolek. soed. **12A**, 282 (1970).
13. Zarev, P., Lipatov, Yu. in: Structure and properties of surface layers of polymers (Russ.), p. 14. Kiev 1972.
14. Malinsky, Yu.: Uspechi chimii **39**, 1511 (1970).
15. Bartenev, G., Zelenev, Yu.: Vysokomolek. soed. **14A**, 998 (1972).
16. Solomko, V.: Thesis. Kiev State University 1971.
17. Houwink, R.: Elastomers & Plastomers. New York 1950.
18. Pod, I., Suchorsky, J., Fischer, A.: Trans. 9th Nat. Vacuum Symp. **1962**, 320.
19. Bills, K., Sweeng, K., Salcedo, F.: J. Appl. Polymer. Sci **4**, 259 (1960).
20. Landell, R.: Trans. Soc. Pheol. **2**, 53 (1958).
21. Gim, A., Pierre, L.: J. Polymer Sci. **B7**, 237 (1969).
22. Baccareda, M., Butta, E.: J. Polymer Sci. **57**, 617 (1962).
23. Kraus, G., Gruver, J.: J. Polymer Sci **A28**, 571 (1970).
24. Galperin, J., Kwei, T.: J. Appl. Polymer Sci. **10**, 617 (1966).
25. Dammont, F., Kwei, T.: Amer. Chem. Soc. Polymer Prepr. **8**, 920 (1967).
26. Kumins, G.: Chem. Canada **13**, 33 (1961).
27. Kumins, G., Roteman, J.: J. Polymer Sci. **A1**, 527 (1963).
28. Lipatov, Yu.: Dokl. Akad. Nayk SSSR **143**, 1142 (1962).
29. Smirnova, A., Pevzner, L.: Dokl. Akad. Nayk SSSR **135**, 663 (1960).
30. Lipatov, Yu.: Dokl. Akad. Nayk BSSR **5**, 69 (1961).
31. Lipatov, Yu., Pavljutchenko, G.: Vysokomolek, soed. **2**, 1564 (1960).
32. Lipatov, Yu., Vasilenko, Ja., Lipatova, T.: Vysokomolek. soed. **5**, 290 (1963).
33. Lipatov, Yu.: Trans. & Plastic Inst. **4**, 83 (1966).
34. Lipatov, Yu.: Plaste u. Kautschuk **N 10**, 738 (1973).
35. Lipatov, Yu.; Vasilenko, Ja.: In: Polymer adhesion (Russ.), Moscow, *1963*, 113.
36. Lipatov, Yu., Sergeeva, L.: Adv. Colloid Interface Sci. **6**, 1, 1976.
37. Lipatov, Yu., Kercha, Yu.: Rubber Chem. Techn. *41*, 537 (1968).
38. Price, E., French, D., Tompa, A.: J. Appl. Polymer Sci **16**, 157 (1972).
39. Sergeeva, L. M., Nesterov, A. E., Lipatov, Yu.: Ukrainian Chem. **41**, 1076 (1975).
40. Babich, V. F., Lipatov, Yu.: unpublished data.
41. Lipatov, Yu., Privalko, V., Kercha, Yu., Mozguchina, L.: Vysokomolek. soed. **13A**, 103 (1971).
42. Droste, D., Dibenedetto, A.: J. Appl. Polymer Sci. **13**, 2149, (1969).
43. Lipatov, Yu., Privalko, V.: Vysokomolek. soed. **14A**, 1648 (1972).
44. Kozlov, P., Timofeeva, V., Kargin, V.: Dokl. Akad. Nauk SSSR **148**, 886 (1963).
45. Lipatov, Yu., Geller, T.: In: Polymer modification (Russ.). p. 66. Kiev 1965.
46. Lipatov, Yu., Geller, T.: Vysokomolek. soed. **8**, 582 (1966).
47. Lipatov, Yu., Geller, T.: Vysokomolek. soed. **9A**, 222 (1967).
48. Alfrey, T., Goldfinger, G., Mark, H.: J. Appl. Phys. **14**, 700, (1943).
49. Michajlov, G.: J. Polymer Sci. **30**, 665 (1958).

50. Michajlov, G.: Makromolek. Chema. **35**, 26 (1960).
51. Slonim, I., Ljubimov, A.: NMR in polymers (Russ.) Moscow 1966.
52. Lipatov, Yu.: Vysokomolek. soed. **7**, 1430 (1965).
53. Lipatov, Yu., Fabuljak, F.: Vysokomolek. soed. **10 A**, 1605 (1968).
54. Lipatov, Yu., Fabuljak, F.: Vysokomolek. soed. **11 A**, 708 (1969).
55. Fabuljak, F., Lipatov, Yu.: Vysokomolek. soed. **12 A**, 738 (1970).
56. Lipatov, Yu.,In: Surface phenomena in polymers (Russ.), p. 7. Kiev 1970.
57. Lipatov, Yu., Fabuljak, F. In: Surface phenomena in polymers (Russ.), p. 15. Kiev 1970.
58. Lipatov, Yu., Fabuljak, F.: J. Appl. Polymer. Sci. **16**, 2131 (1972).
59. Lipatov, Yu., Fabuljak, F. In: Mechanism of relaxation phenomena (Russ.), p. 274. Moscow 1972.
60. Lipatov, Yu., Fabuljak, F.: Vysokomolek. soed. **15 A**, 1513 (1973).
61. Lipatov, Yu., Sergeeva, L., Todosijchuk, T.: Dokl. Akad. Nauk SSSR **218**, 1144 (1974).
62. Cole, K., Cole, R.: J. Chem. Phys. **2**, 341 (1941).
63. Cole, R., Davidson, D.: J. Chem. Phys. **20**, 1389 (1952).
64. Zelenev, Yu., Bartenev, G.: Ukr. Phys. J. **12**, 888 (1967).
65. Zelenev, Yu., Molotkov, A.: Vysokomolek. soed. **6**, 1426 (1964).
66. Lipatov, Yu. In: Fillers for polymer, (Russ.), p. 9. Moscow 1969.
67. Lipatov, Yu., Fabuljak, F., Kuznezova, V.: Voprosy chimii i chimicheskoy technologii, N 32, p. 3. Hkarkov 1974.
68. Lipatov, Yu., Fabuljak, F.,Watamanjuk, V.: Dokl. Akad. Nayk SSSR **212**, 925 (1973).
69. Zelenev, Yu.: Thesis, Moscow 1971.
70. Brown, V.: Dielectrics, p. 265. Moscow 1961.
71. Solomko, V.: Dokl. Akad. Nayk USSR, B, N **9**, 815, 19169.
72. Zelenev, Yu., Kardanov, H., Solomko, V.: Plaste u. Kautschik **19**, 263 (1972).
73. Lipatov, Yu., Fabuljak, F.: Sintez i phisiko-chimia polymerov, N 17, 18–22. Kiev 1975.
74. Fabuljak, F., Lipatov, Yu.: Yysokomolek. soed. V **12**, 871 (1970).
75. Lipatov, Yu.,Fabuljak, F.,Sergeeva, L.: Dokl. Akad. Nayk Ukr. SSR, B. **1974**, 551.
76. Waldrop, A., Krauss, G.: Rubb. Chem. Techn. **42**, 1155 (1969).
77. Fudzimoto, K.,Hilsi, T.: J. Soc. Rubber Ind. Japan **43**, 88 (1970).
78. Ilivividze, V., Klimanov, S., Leznev, I.: Vysokomolek. soed. **A9**, 1924 (1967).
79. Lipatov, Yu., Fabuljak, F., Shifrin, V.: Vysokomolek. soed. **A18**, 763 (1976).
80. Lipatov, Yu., Fabuljak, F.: Dokl. Akad. Nayk SSSR **205**, 635 (1972).
81. Lipatov, Yu., Fabuljak, F.: Vysokomolek. soed. **A15**, 1272 (1973).
82. Lipatov, Yu., Fabuljak, F., Ovchinnikova, G. In: Physical Chemistry of Polymer Compositions (Russ.), p. 74. Kiev 1974.
83. Lipatov, Yu. S., Fabuljak, F., Shifrin, V.: Vysokomolek. soed. **A18**, 767 (1976).
84. Smallwood, H.: J. Appl. Phys. **15**, 758 (1944).
85. Guth, E.: O. Cold. Phys. Rew. **53**, 322 (1938).
86. Guth, E.: Rubb. Chem. Techn. **18**, 596 (1945); J. Appl. Phys. **16**, 20 (1945).
87. Eilers, H.: Koll. Z. **97**, 313 (1941).
88. Kerner, E.: Proc. Roy. Soc. **69B**, 808 (1956).
89. Tsai, S.: Formulas for the Elastic Properties of Fiber-reinforced Composites. AD 834851, June 1958.
90. Lewis, T., Nielsen, L.: J. Appl. Polymer Sci. **14**, 1449 (1970).
91. Ziegel, K., Romanov, A.: J. Appl. Polymer Sci. **17**, 1119 (1973).
92. Zgaevsky, V.: Thesis, Leningrad Institute of Textile Industry 1970.
93. Zgaevsky, V., Galizky, Yu.: Referativn. Zurnal Chimia **10**, C 165 (1973).
94. Nielsen, L.: J. Appl. Phys. **41**, 4626 (1970).
95. Nakagama, J., Takenchi, R.: Mem. Inst. Sci. a. Res. Osaka Univ. **30**, 75 (1973).
96. Dickie, R.: J. Appl. Polymer. Sci. **17**, 2509 (1973).
97. Bueche, A.: J. Polymer Sci. **25**, 139 (1957).
98. Parkinson, D.: Reinforcement of Rubber. London 1957.
99. Houwink, R.: H. Janssen, Rubb. Chem. Techn. **29**, 409 (1956).
100. Rivlin, R., Thomas, A.: J. Polymer Sci. **10**, 291 (1953).

101. Payne, A. In: Reinforcement of elastomers (G. Kraus). Interscience Publ.
102. Nielsen, L., Lewis, T.: J. Polymer Sci. **A2**, 7, 1705 (1969).
103. Lipatov, Yu., Babich, V., Rosovizky, V.: J. Appl. Polymer Sci. **18**, (1213).
104. Lipatov, Yu., Babich, V., Korzuk, N.: Vysokomolek. soed. **A16**, 1629 (1974).
105. Ferry, J.: Viscoelastic properties of polymers. New York 1961.
106. Kuznezov, G., Irgen, L.: Mechanika polymerov. **3**, 487 (1973).
107. Sagalaev, G., Simonov-Emeljanov, I., Babkova, L.: Plastmassy, **N 2**, 51 (1974).
108. Simonov-Emeljanov, N.: Thesis, Moscow Lomonosow Institut, 1973.
109. Sagalaev, G., Simonov-Emeljanov, I.: Plastmassy N **2**, 48 (1973).
110. Kuleznev, V. N.: Thesis, Institute of Fine Chemical Technology. Moscow 1973.
111. Vojuzky, S., Zayonchkovsky, A., Kargin, V.: Dokl. Akad. Nayk SSSR **94**, 1093 (1954).
112. Slonimsky, G., Musaeljan, I., Kazanzeva, V.: Vysokomolek. soed. **6**, 219 (1964).
113. Slonimsky, G., Musaeljan, I., Kazanzeva, V.: Vysokomolek. soed. **6**, 818 (1964).
114. Takayanagi, M., Nemura, S., Minami, S.: J. Polymer. Sci. **C5**, 113 (1964).
115. Krauss, G., Rollmann, K.: Macromolecules **3**, 92 (1970).
116. Krauss, G., Rollmann, K.: Multicomponent polymer systems (F. Gould ed.), Amer. Chem. Soc.
117. Dickie, P.: J. Appl. Polymer Sci. **17**, 45 (1973).
118, Dickie, P., Cheung, M.: J. Appl. Polym. Sci. **17**, 79 (1973).
119. Pelzbauer, Z.: Plaste u. Kautschuk **20**, 748 (1973).
120. Wang, T., Schonhorn, H.: J. Appl. Phys. **40**, 5131 (1969).
121. Bueche, F.: Physical properties of polymers. J. Wiley: New York 1962.
122. Becker, G., Oberst, H.: Koll Z. Z. Polymere **148**, 6 (1956).
123. Blatz, P.: Ind. Eng. Chem. **48**, 727 (1956).
124. Payne, A. In: Rheology of elastomers (P. Mason ed) London 1958.
125. Ecker, R.: Kautschuk u. Gummi **21**, 304 (1968).
126. Lewis, T., Nielsen, L.: J. Appl. Polymer. Sci. **14**, 1449 (1970).
127. Landell, R.: Trans. Soc. Rheol. **2**, 53 (1958).
128. Robinson, J.: Trans. Soc. Rheology **1**, 18 (1957).
129. Bueche, A.: J. Appl. Polymer. Sci. **5**, 271 (1961).
130. Bartenev, G. M., Vishnizkaya, L.: Kolloidny journal **18**, 135 (1956).
131. Ljalina, N., Zelenev, Yu., Bartenev, G.: Vysokomolek. soed. **B10**, 510 (1968).
132. Ljalina, N., Zelenev, Yu., Bartenev, G. In: Mechanismy relaxazionnych javlenij. Nauka: Moscow 1972.
133. Bartenev, G., Ljalina, N.: Vysokomolek. soed. **A12**, 922 (1970).
134. Ordgonikidze, S., Margoli, A., Pochil, P.: Mechanika polymerov,**N4**, 688 (1968).
135. Leznev, N., Jampolsky, B., Ljalina, N., Volodina, V.: Kautschik i rezina **2**, 16 (1965).
136. Leznev, N., Jampolsky, B.; Dokl. Akad. Nayk SSSR **160**, 861 (1965).
137. Ljalina, N., Leznev, N., Zelenev, Yu., Bartenev, G. In: Mechanismy relaxazionnych javlenij, p. 295, 1972.
138. Sato, Y.: J. Soc. Rubber Ind. Japan **39**, 185 (1966).
139. Albrecht, B., Freundentahl, A.: Rheol. Acta **1**, 431 (1961).
140. Zelinger, J., Heidingsfeld, V.: Sd. Vysoké Skoly Chem. techn. Prace **C 12**, 71 (1967).
141. Zelinger, J.: Intern. J. Polymeric Mater. **1**, 317 (1972).
142. Sierakowski, R., Nevill, G., Howse, D.: J. Composite Mater. **5**, 118 (1971).
143. Jenovsky, Yu., Frenkin, E., Vinogradov, G.: Mechanika polymerov **N4**, 752 (1968).
144. Vovkotrub, N., Surovzev, V., Solomko, V.: Vestnik KGU ser. chim. **N11**, 53 (1970).
145. Solomko, V., Surovzev, V., Pasko, S. In: Mechanismy relaxazionnyc̆h javlenij, p. 238, 1972.
146. Kotenkov, V., Guzeev, V., Malinsky, Yu. In: Structure and properties of surface layers of polymers (Russ.), p. 179. Kiev 1972.
147. Molotkov, A., Gunkin, S., Zelenev, Yu. In: Structure and properties of surface layers of polymers (Russ.), p. 179. Kiev 1972.
148. Freidin, A., Kiem, Vu. Ba., Zelenev, Yu. In: Structure and properties of sorface layers of polymers (Russ.), p. 226. Kiev 1972.
149. Solomko, V., Surovzev, V. In: Mechanizmy relaxazionnyc̆h javlenij, p. 238. Moscow 1972.

150. Ziegel, K., Romanov, A.: J. Appl. Polymer Sci. **17,** 1133 (1973).
151. Zelenev, Yu., Elektrova, L.: Vysokomolek. soed. **A11,** 2123 (1973).
152. Zelenev, Yu., Aivazov, A.: Mechanika polymerov **6,** 138 (1970).
153. Mizumachi, H., Fujino, M.: Holzforschung **26,** 154 (1972).
154. Lipatova, T., Myschko, V. In: Physical chemistry of polymer compositions (Russ.), p. 90. Kiev 1974.
155. Klempner, D., Frisch, H., Frisch, K.: J. Elastoplastics **3,** 2 (1971); J. Polymer Sci. **8,** A-2, 921 (1970).
156. Lipatov, Yu., Sergeeva, L.: Uspechi chimii **45,** 138 (1976).
157. Lipatov, Yu., Babich, V., Karabanova, L., Korzuk, N., Sergeeva, L.: Dokl. Akad. Nauk Ukr. SSR **B 1,** 39 (1976).
158. Lipatov, Yu., Kercha, Yu.: in press.
159. Lipatov, Yu., Babich, V.: in press.
160. Relaxation Phenomena in Polymers (Russ.), p. 187. Chimica: Leningrad 1972.
161. Lipatov, Yu., Babich, V.: to be published.
162. Babich, V. In: New methods of polymer investigating (Russ.), p. 118. Kiev 1975.
163. Lebedev, E., Lipatov, Yu., Bezruk, V.:In: New methods of polymer investigating (Russ.), p. p. 3. Kiev 1975.
164. Lipatov, Yu., Rosovizky, V., Babich, V.: Vysokomolek. soed. **B 16,** 512 (1974).
165. Ziegel, K., Frensdorf, H., Fogiel, A.: J. Appl. Polymer Sci. **13,** 867 (1969).
166. Gurevich, G.: Deformability of media and propagation of seismic waves (Russ), Nauka, Moscow, 1974.
167. Martin, D.: Proc. Army sci. Cont. West Point, New York 1964, Vol. 2; Washington D. C. 1965, p. 211–226.
168. Hachiu, Z.: Trans. Soc. Rheol. **1965,** 9.
169. Zgaevsky, V., Frenkel, S., Zelenev, Yu. In: Structure and properties of surface layers of polymers (Russ.), p. 147. Kiev 1972.
170. Dimitrova, D., Aivazov, A., Zelenev, Yu.: Godichnik Vysch. chim.-technol. inst. Burgas **8,** 57, 1972 (1971).
171. Kaplan, D., Tschoegl. N.: Amer. Chem. Soc. Polym. Prepr. **14,** 1090 (1973).

Received May 17, 1976

Electro-Optic Methods for Characterising Macromolecules in Dilute Solution

Barry R. Jennings
Physics Department, Brunel University, Uxbridge, Middlesex, U. K.

Table of Contents

A brief survey is presented of a number of methods which are used to characterise the optical, electrical and geometrical properties of macromolecules in dilute solution. In the presence of an electric field, molecular order is introduced with resulting changes in the optical properties of the solution as a whole. Principles of the methods and illustrative data are given for electrically induced birefringence, dichroism, optical rotation and light scattering. The methods are not only able to lead to values of size, shape and dipole moments of molecules, but also to give discrete polydispersity averages for rigid particles and to indicate molecular flexibility. The methods are extremely rapid. Finally some novel experiments, in which the electric field associated with a laser beam is used to impose order and hence induce electro-optic effects, are mentioned.

I. Introduction

The optical properties of a material have long been recognised as characteristics of the structure of that material. The information is especially of value when crystalline or highly ordered samples are studied as one can relate the optical functions directly to the atomic or molecular array. With solutions or suspensions of macromolecules, order is generally only of very short range when compared with the molecular dimensions, as the molecules themselves adopt a random arrangement within the bulk of the medium. It is for this reason that optical methods are generally restricted to intramolecular information for such systems.

Currently, there is a growing interest in novel electro-optical methods for the study of polymer and biopolymer solutions. Briefly stated, the principle is as follows. The majority of polymers, biopolymers, viruses and colloidal particles are anisotropic in their optical properties. It is easy to visualise that the bonding along a polymer backbone is different in type and density to that across such an axis. Light beams polarised parallel and perpendicular to this axis will interact differently with the polymer. A typical example is the speed with which each beam may traverse the molecule; a property generally referred to through the refractive index (n) of the material. Hence, polymers are generally anisotropic in their refractive index (Fig. 1). In dilute solution, owing to the random molecular array, this directional property is not seen in the behaviour of the solution as a whole. However, anisotropic bonding generally also results in electrical anisotropy. This may result from either fixed charge separations giving rise to permanent dipole moments (μ) or from the relative ease with which charges can be separated in the presence of an electric field. This last mentioned arises from the variation of the polarisability (α) with molecular direction and gives rise to induced dipole moments. Hence, in an electric field, the solute molecules can be caused to partially align and thereby impose their optical anisotropy upon the medium as a whole. Measuring the changes in a given optical property of the solution simultaneous with the application of the electric field can lead to the following molecular information. Firstly, the basic optical parameters associated with each molecular axis can be evaluated without the need to concentrate and crystallise the material. Secondly, values of the electrical parameters μ and α can

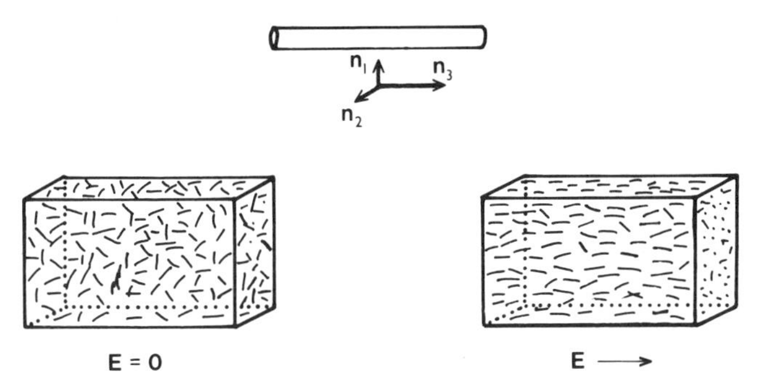

Fig. 1. Schematic representation as to how solute order imposes the molecular anisotropy on the bulk solution. The example utilises anisotropy of refractive indices (n_i) of rod molecules in an electric field E

be determined and related to the structure of the molecules. Thirdly, by observing the *rates* at which the optical properties change in response to the electric field, we have a direct measure of the rotary dynamics of the molecules as they exhibit free motion in their viscose environment. This is characterised through a rotary relaxation time (τ) which can be related directly to the size and shape of the macromolecules.

In this brief survey the author will outline a number of electrooptical methods in current use and give illustrative results which demonstrate their practical utility. Theory for the various effects will be given predominantly for cylindrically symmetric molecules which are rigid, uncharged and non-interacting as this is the best developed to date and allows a useful comparison between the methods. Reference will be made to work done on flexible systems. The object of this survey is primarily to introduce those scientists who are on the look-out for novel methods to characterise macromolecules in dilute solution to electro-optic experiments and to demonstrate the potential and versatility of the methods. The selected illustrative data on various polymer solutions are not intended to be comprehensive. They have simply been drawn from studies conducted in the author's research group.

II. Electric Birefringence

Commonly known as the Kerr Effect, this is the best known electro-optic phenomenon. Although initially studied in glasses by John Kerr (*1*) in 1875, who considered the birefringence to be related to electrically induced strain in the material, it is now used widely to follow the alignment due to orientation and deformation of macroparticles in solution and suspension (*2–4*). It owes its origin to anistropy of the refractive indices associated with the major geometric axes of the molecules.

The effect can be manifest by placing the polymer solution in a suitable cell (Fig. 6a) between a pair of electrodes. If a beam of well collimated light is passed centrally through this cell, with the light initially linearly polarised at 45° azimuth to the electric field direction, then elliptically polarised light will leave the cell when the solution is made birefringent. This will only be when the electric field is applied across the cell. A quarter wave plate (Q) and analysing polariser (A) are set in parallel azimuth to each other, but are 'crossed' with the initial state of linear polarisation. It is convenient to record the light penetrating such a system using a photomultiplier and oscilloscope as indicated in Fig. 2. A low power laser gives a convenient, well collimated light beam for the apparatus. However, it does restrict the

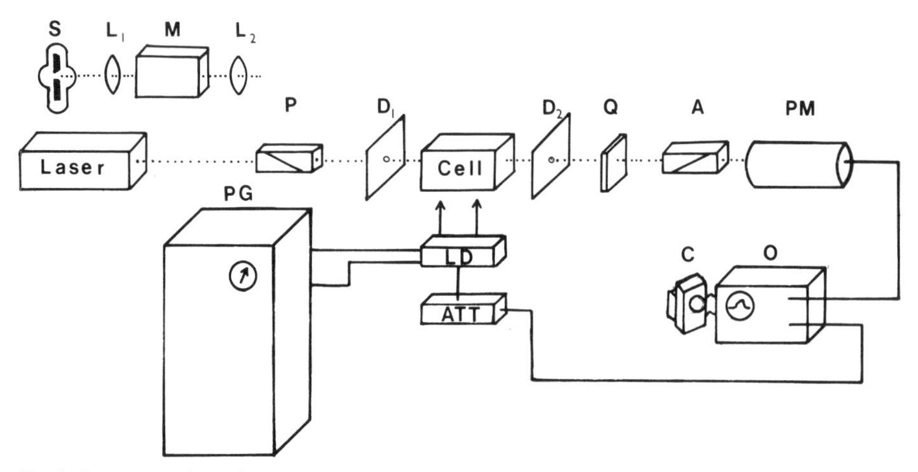

Fig. 2. Representation of an apparatus.
The laser, or the combination of the arc (S) lenses (L) and monochromator (M), forms the light source. Other optical components are P – polariser, D – diaphragms, Q – quarter-wave plate, A – analyser. The electronic components are PM – photomultiplier, O – oscilloscope, C – camera, PG – pulse generator, LD – load and ATT – attenuator

wavelengths available. A reasonable alternative is an arc source provided that it is powered from a suitably smoothed supply. Details of suitable optical components and their commercial availability have been given by the author and a colleague elsewhere (*5*), where the method of obtaining Δn from the photomultiplier signal is also described.

The birefringence (Δn) observed in the solution is defined as the difference in the refractive indices of the solution parallel ($n_{\parallel}$) and perpendicular ($n_{\perp}$) to the electric field direction. It is given in terms of the optical polarisability (g) anisotropy associated with the solute molecules, their volume concentration C_v and the average refractive index of the solution (n) by the expression (*6*)

$$\Delta n = (n_{\parallel} - n_{\perp}) = (2\,\pi\, C_v/n)\,(g_3 - g_1)\,\Phi \qquad (1)$$

Here the subscripts 3 and 1 indicate the relevant parameters along the major (3) and a minor (1) axis of a cylindrical molecule or segment respectively. The orientation factor Φ describes the average orientation of the molecules at a given time (t) or in a given electric field of amplitude E. Should this be an alternating electric field, a root mean square value for E is implied. The function Φ has been tabulated and computed for a variety of conditions (*2*). We shall restrict ourselves to low degrees of orientation as when the electrical torque on the molecules is much smaller than the thermal energy (kT). Such low fields are unlikely to interfere with the molecular structure.

With rigid molecules for a sinusoidal electric field of angular frequency ω,

$$\Phi = \left(\frac{(\mu_3^2 - \mu_1^2)}{(1 + \omega^2/4D^2)\, k^2T^2} + \frac{(\alpha_3 - \alpha_1)}{kT} \right) E^2/15 \tag{2}$$

The parameters μ_i and α_i are the magnitudes of the permanent dipole moments and the electrical polarisabilities respectively. The factor Φ is independent of the optical property being measured. It simply describes the interaction of the applied field with the electrical characteristics of the molecules. Because of this it will be recurrent in the theoretical description of the other electro-optical phenomena described below. Before returning to electric birefringence we note a few points from Eq. (2). Firstly, if we evaluate Φ from a particular experiment, its quadratic dependence on E will indicate the appropriateness of the experimental conditions (i. e. suitably low field strength) and enable one to isolate the electrical factor within the brace of the equation. Secondly, only the μ term is significantly dependent on the frequency ω. Hence by analogy with measurements of the real part of the relative permittivity in dielectric dispersion experiments, the electro-optical effects will exhibit frequency dispersion properties. By making measurements as a function of ω, we shall be able to isolate the permanent and induced dipolar contributions. Thirdly, the parameter D is the rotary diffusion constant of the molecules, often related to a rotary relaxation time ($\tau = {}^1/_{6D}$). It can be readily evaluated from a frequency dispersion curve by finding that critical frequency (f_c) at which the dispersion curve amplitude has suffered half its total change, whence

$$D = \pi f_c \tag{3}$$

equations have been given in the literature which relate D to the length L of rigid rods (*7*), the diameter of a disc (*8*) or the axial ratio of rigid ellipsoids (*8*). Fourthly, it is convenient and advantageous to apply the electric field in the form of a short duration pulse rather than continuously. In this way, the methods are very fast and problems due to heating effects and electrophoresis are minimised. The pulse duration must be long enough for the solute molecules to reach an equilibrium condition of orientation before the field is terminated. Such orientation in a resistive medium requires a finite time as does the disorientation process after the passing of the pulse. These rates will be reflected in the transient changes in the birefringence as recorded on the oscilloscope (Fig. 2). Hence rather than apply many fields to the solutions and obtain a frequency dispersion curve of the optical changes, it is convenient to apply two successive pulses and record the accompanying optical changes. These pulses would conveniently be of a dc field followed by a high frequency sinusoidal field of such frequency as to greatly exceed the critical frequency (Fig. 3). A comparison of the amplitudes of the optical responses leads directly to the ratio $(\mu_3^2 - \mu_1^2)/(\alpha_3 - \alpha_1)\, kT$. Absolute values of the permanent dipole moment and the polarisability difference can be obtained using the auxilliary data implied in Eq. (1).

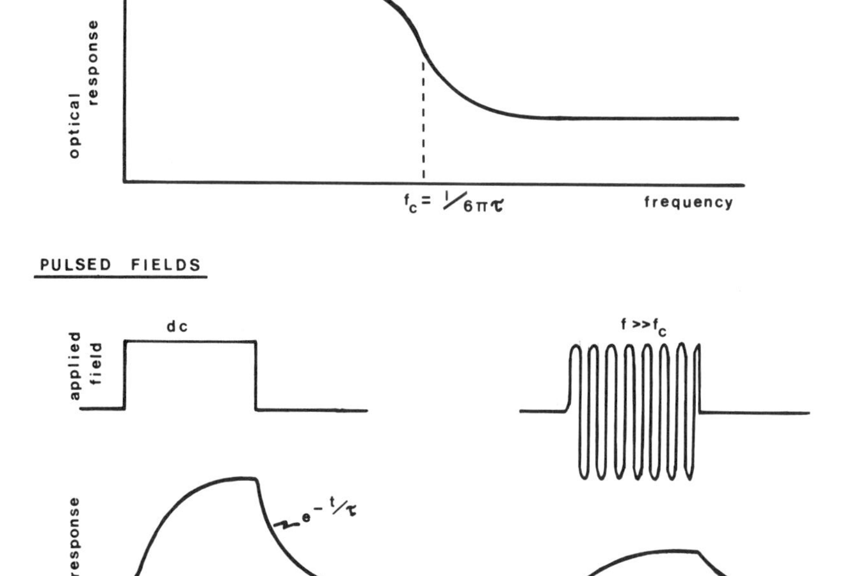

Fig. 3. Schematic frequency dispersion of an electro-optic response. Top figure represents a continuous dispersion. Lower figure indicates how two pulsed measurements can be used. The value of f_c has been based on the rotation of rigid molecules

A major advantage in using pulsed fields is the rapid evaluation of D and τ from the rate of decay of the optical response following the field. For a monodisperse solution, the disorientation process is a simple first order rate process and therefore obeys the equation

$$\Delta Z = \Delta Z_0 \exp(-t/\tau) \tag{4}$$

where ΔZ is the amplitude of the change of the optical property at any time t after the cessation of the pulse. At $t = 0$, the amplitude is ΔZ_0. With birefringence, ΔZ is replaced by Δn. A good estimate of τ and hence the particle size can be made simply by observing the oscilloscope trace and noting that time taken for the transient to decay to approximately a third of its maximum value.

The results of a study on poly-(n-butyl isocyanate) in benzene illustrate the method. Four samples of different molecular weight in the range $2 \times 10^4 < \overline{M}_W < 3 \times 10^5$ were studied. For fields up to 24 kV cm^{-1}, the amplitude of the birefringence was still in the region defined by Eqs. (1) and (2) where Δn was proportional to E^2. A typical transient as recorded on the oscilloscope is shown in Fig. 4 alongside a frequency dispersion plot of the normalised birefringence amplitude. From these results we note the following information. Firstly, the dispersion curve indicates that only permanent dipoles play a significant role and that the polarisability con-

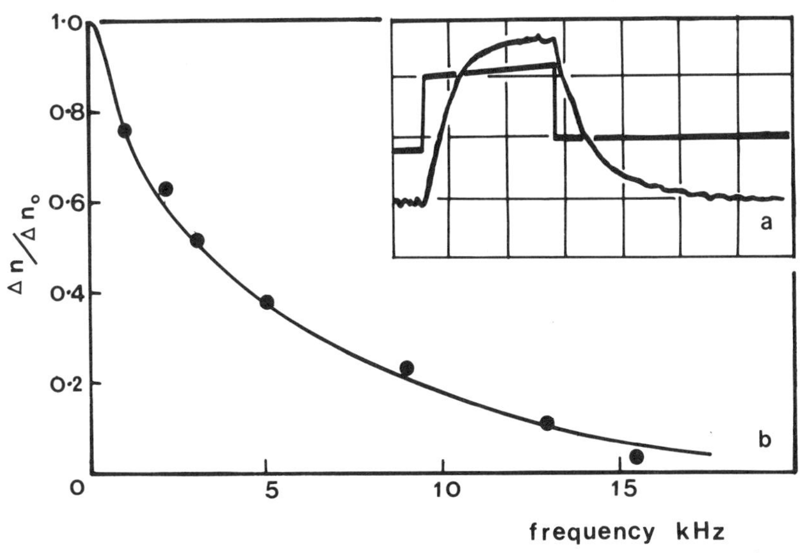

Fig. 4. Electric birefringence of polyisocyanate in benzene. The transient (a) was for a sample with $\overline{M}_W = 30 \times 10^4$, in a field of 24 kV cm^{-1} amplitude and 25 μs duration. The dispersion curve (b) was for $\overline{M}_W = 13 \times 10^4$, in a field of 8 kV cm^{-1}. For both samples, $c = 2.5 \times 10^{-3}$ g ml^{-1}. Data from Ref. (*10*)

tribution is negligible. Secondly, the positive nature of the permanent dipole contribution indicates that $\mu_3 \gg \mu_1$ and that such a moment is predominantly directed *along* the major molecular axis. Thirdly, the value of the rotary diffusion constant was $1.57 \times 10^4 s^{-1}$ from the critical frequency and $1.66 \times 10^4 s^{-1}$ from the decay of the relevant transient, thereby justifying the theory in the case of non polar solvents.

Ever since their synthesis, it has been conjectured that the poly-(isocyanates) exist in solution as highly stiffened molecules, probably as shallow helices (*9*). By studying samples of various molecular weights and plotting values of D or τ as suitable functions of $\overline{M}$ one can see that whereas the rigid rod model is not applicable for all M, it is increasingly appropriate at the lower molecular weights. Suitable extrapolation of the data to low molecular weight enables one to isolate two useful parameters namely, μ_0 and L_0. These are the monomer axial dipole moment and the length of a monomer projected along the helix axis respectively. Full details of the evaluation of these parameters can be found elsewhere (*10*). The value of $L_0 = 0.13$ nm is less than the monomer length and hence indicates that the molecule winds into a helix. Furthermore, the dipole moment has its origin predominantly in the C = 0 bond. As the resultant molecular dipole moment is along the helix, then the molecule is not only helical but adopts the cis configuration (Fig. 5), thereby confirming earlier dielectric data (*11*).

Finally, what of the higher molecular weights? Increasing molecular weight for such a shallow helix must result in increasing flexibility. Hearst (*12*) has presented equations for D in terms of the 'persistence length' (q) of a weakly bending rod and of a worm-like chain. The parameter q is the projection of the molecule as if it were

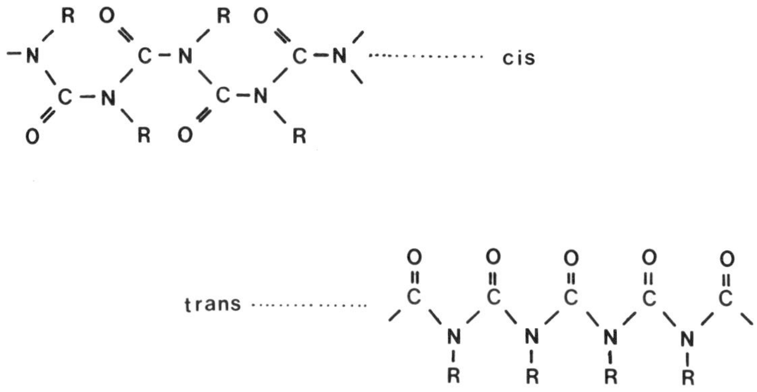

Fig. 5. Possible configurations of polyisocyanate. The dipole moment determination indicates that the cis form exists in benzene solutions. Here R represents C_4H_9

of infinite length, on to the direction of the first bond. It has the values of ∞ to zero for rods or completely flexible chains respectively. Using these equations, it could be seen that a value of q of the order of a few hundred was indicated by the birefringence decay data for sufficiently large M.

For a wider variety of examples on proteins, polyelectrolytes and other polymers, the reader is referred to fuller review articles (*2–4*).

III. Electric Linear Dichroism

Refractive index is a complex property, generally expressed in the form

$$n(1 - j\kappa) \tag{5}$$

Hitherto we have been concerned with molecules exhibiting anisotropy in the real part of the refractive index. Such molecules logically would be expected to have anisotropic imaginary components; but the imaginary part of the refractive index represents energy loss and is related to the optical absorption of a material. Thus, if a beam of incident intensity I^* passes through a cell of length 1 containing a solution of concentration c and the transmitted intensity be I, then, according to the Beer-Lambert law

$$I = I^* \exp(-\epsilon c l) \tag{6}$$

where ϵ is a molecular extinction coefficient. It is related to κ, the index of absorption, through the form $\epsilon = 4\pi\kappa/c\lambda$, where c is the vacuum velocity of light. A macromolecule can be expected, in those regions of the electromagnetic spectrum where absorption is experienced, to have various extinction coefficients ϵ_i associated with incident light which is polarised parallel to the various molecular axes. In the absence of an electric field, a solution of such molecules will absorb light according to the average extinction coefficient for the molecule. On molecular alignment within an applied field, the solution will manifest changes in its adsorptive power. It is then termed 'dichroic'. Hitherto it has been conventional to measure such changes for the two specific cases where the initial light beam is linearly polarised first parallel to and then perpendicular to the applied field vector. A cell such as that of Fig. 6a

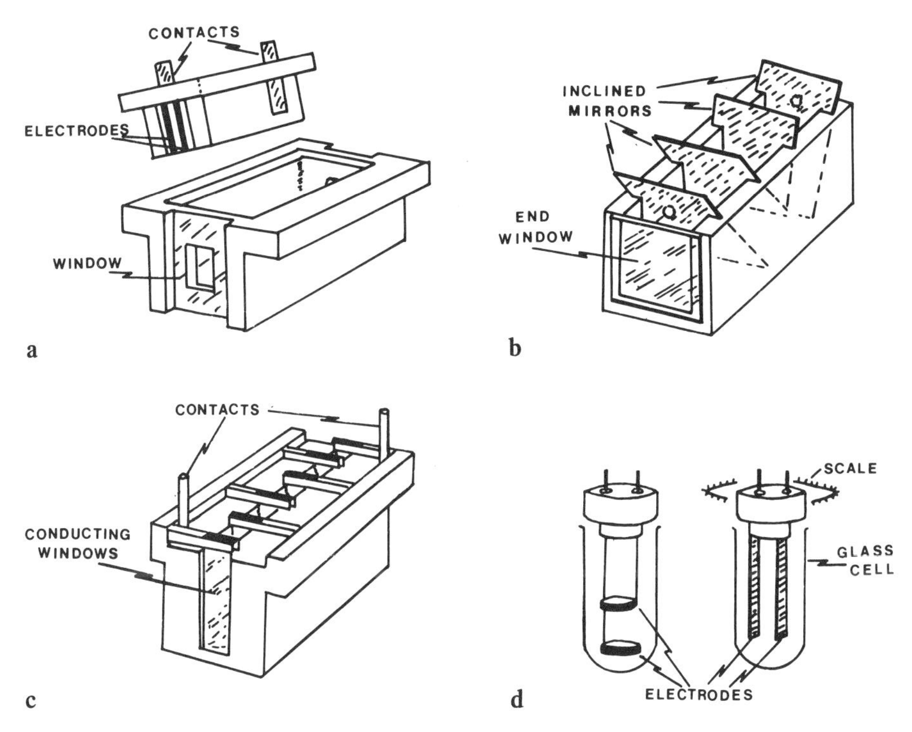

Fig. 6. Cells used in various electro-optic experiments. In (a) the field is transverse to the light path as for birefringence and dichroism experiments. (b) is the multireflecting cell for electric dichroism. In (c) the multiple cells are also arranged in longitudinal array for optical rotation experiments. (d) are light scattering cells with horizontal and vertical electrodes as shown

would be used; the spectral wavelength would be chosen using the monochromator of Fig. 2 and the analyser (A) and quarter-wave plate (Q) of that figure would be removed. In all other respects, the apparatus would be as described in the previous section.

In the case of rigid cylindrically symmetrical molecules, we have recently shown in this research group (*13*) that a single transient measurement is all that is required if the field be applied collinear with the light beam. For such longitudinal measurements, the polariser can also be dispensed with and the measurements made in a modified spectrophotometer (*14*). A special cell is then required in which the field can be applied along the optical path. Initially, this was achieved by coating the inner faces of the entrance and exit windows of a spectrophotometer cell with thin, conducting metal oxide films. More recently a multiple reflecting cell has been designed (*15*) in which the light beam is reflected back and fore between two conducting mirrors which also act as the electrodes. Such a cell gives a much greater optical path length and hence increased sensitivity to the method.

If we use $\epsilon^{\parallel}$ and $\epsilon^{\perp}$ for the extinction coefficients observed with the initial polarisation state parallel or perpendicular to the applied field and use ϵ^{L} for the longitudinal situation, then in the electric field

$$\left.\begin{aligned} \epsilon^{\parallel} &= \epsilon_u + \frac{2}{3}(\epsilon_3 - \epsilon_1)\,\Phi \\ \epsilon^{\perp} = \epsilon^{L} &= \epsilon_u - \frac{1}{3}(\epsilon_3 - \epsilon_1)\,\Phi \end{aligned}\right\} \qquad (7)$$

where $\epsilon_u = (\epsilon_3 + 2\epsilon_1)/3$ and is the coefficient recorded in the absence of any field. Hence, if one knows or can otherwise determine Φ, discrete values can be found for ϵ_3 and ϵ_1 at any particular wavelength. It is particularly useful to realise that if one can generate fields of sufficiently high amplitude so that complete molecular alignment can be obtained, then Φ equals unity and the above mentioned values of ϵ_i can be realised. Measurements can then be repeated at low field strength where $\Delta\epsilon$ is proportional to E^2 (where $\Delta\epsilon$ represents the difference between $\epsilon^{\parallel}$, $\epsilon^{\perp}$ or ϵ^{L} and ϵ_u). In these cases Φ has the form of Eq. (2) and the dipolar properties can be found as described in the section on electric birefringence. In addition, values of τ and D can be obtained from the decay of the transient absorption changes as $\Delta\epsilon$ can be substituted for ΔZ in Eq. (4).

An interesting and illustrative example is provided by a study on a dilute aqueous suspension of copper phthalocyanine crystallites. These needle-like particles are composed of layers of regularly stacked planar molecules. According to X-ray data (*16*) the molecular 'plates' adopt a continuous zig-zag configuration. The crystallite suspensions exhibit characteristic spectra in the red region of the visible spectrum. A strong absorption line occurs at 720 nm. Transient electric dichroism measurements were made at this wavelength on dispersions of approximately 4×10^{-6} g ml^{-1} concentration, for electric fields of up to 6.5 kV cm^{-1}. In Fig. 7 it is seen that such fields are sufficient to cause complete orientation saturation. A single transient at this high field condition was sufficient to enable the extinction coefficients ϵ_3 and ϵ_1 to be evaluated. From the transients at low fields, that is within the region of a quadratic dependence on E, the decay rate corresponded to an average particle length of 130 nm which was in good agreement with electron microscopic data for this sample. In addition, no permanent dipole moment was evident. A polarisability difference of 8.5×10^{-25} F m^2 per unit crystallite length was obtained.

These experiments were then repeated at a number of wavelengths in the spectral range $500 < \lambda$ (nm) < 750. In each case ϵ_u, ϵ_3 and ϵ_1 were determined. Figure 8 displays these parameters and indicates some interesting facts. The conventional spectrum as obtained in the absence of any electric field is that shown for ϵ_u. However, by reducing this to its composite parts ϵ_3 and ϵ_1 we note that, the peak at

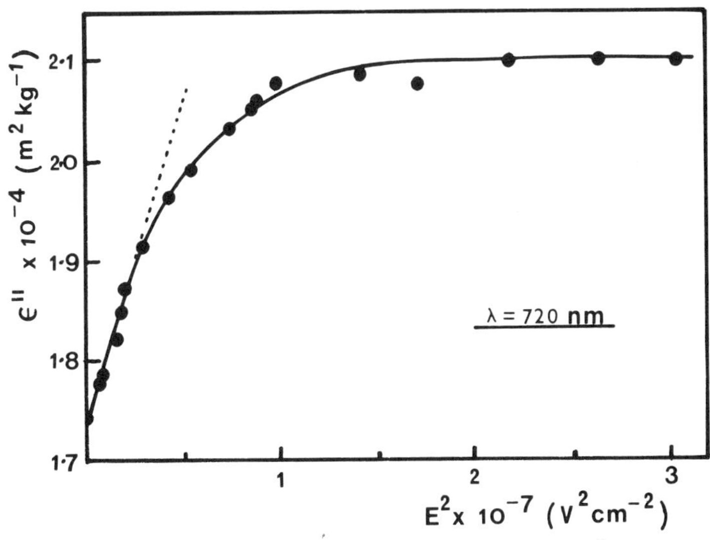

Fig. 7. Field strength dependence of the extinction coefficient $\epsilon^{\parallel}$ for copper phthalocyanine in water at 720 nm wavelength; $c = 3 \cdot 5 \times 10^{-6}$ g ml^{-1}.
Data from Ref. (*17*)

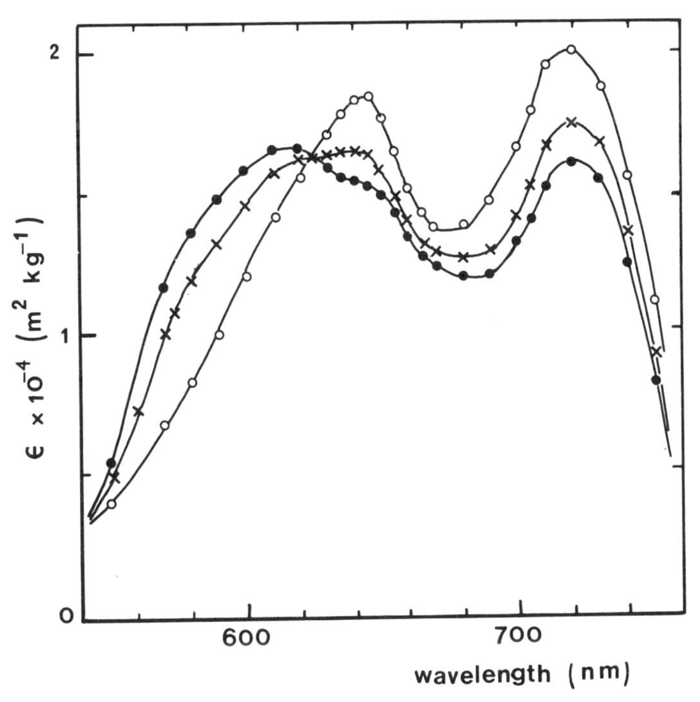

Fig. 8. Spectral variation of the extinction coefficients for copper phthalocyanine crystallites in water. Crosses, filled circles and open circles represent ϵ_u, ϵ_1 and ϵ_3 respectively for $c = 4 \times 10^{-6}$g ml^{-1}. Data from Ref. (*17*)

720 nm is due to a transition moment directed predominantly *along* the crystallite axis. This is also true for the peak in the region of 640 nm. At lower wavelengths the variation ϵ_u indicates the probable existence of another peak in the region of 580 nm. The spectra of ϵ_3 and ϵ_1 indicate that such a peak must exist as these curves change over. Any such peak must then have its associated transition moment predominantly directed *across* the crystallite axis. The spectra were analysed so as to isolate the three supposedly Gaussian profiles: The composite peaks were found to be at 720, 640 and 578 nm. In addition, the specific values of ϵ_3 and ϵ_1 at each of these wavelengths had their origins in specific transition moments directed at specific directions with respect to the crystallite axis. In the present study it was not only possible to estimate the direction of these transition moments within the crystallites but also to relate them to atomic origins within the planar molecules which constitute the crystallites. Full details of the results and the analysis for this particular study can be found in the original publication (*17*). To the author's knowledge, the only alternative procedure for obtaining such molecular information is to grow large crystals of the material in question and to study their optical properties. This is not always convenient. When it is remembered that these electric dichroism studies were made on extremely dilute suspensions of the crystallites and that the size and electrical polarisability were also obtained, the value of such studies can be appreciated.

Finally, it is pointed out that, with coloured solutions and suspensions, transient electric dichroism measurements can be used instead of electric birefringence to determine τ and hence particle size for solutions and suspensions. These τ values can be analysed to indicate particle flexibility in the same manner as was described above for the poly-(isocyanates).

IV. Electro-Optical Rotation

The ability of a solution of macromolecules to rotate the plane of polarisation of a transmitted light beam is often used both to measure the concentration of solutions and as an empirical guide to local conformation changes. The optical rotary power reflects the local order of the molecules. It gives little information on their gross size. If a solution of concentration c causes an observed rotation of θ degrees through a cell of length 1, then the specific rotation $[\theta]$ is given by

$$[\theta] = \theta/cl. \tag{8}$$

What is not generally appreciated is that the phenomenon has its origins in certain specific electronic transitions within the molecular structure. The optical rotation is thus another entity which has different values associated with the various axes of a molecule. Fresnel (*18*) originally related optical activity to helical symmetry in a structure and this led to the well known test of optically active molecules, namely that if the structure was asymmetric, (*i.e.* its mirror image was not superposable on itself), the molecules would be optically active. The term 'chirality' has also been applied to this type of molecular symmetry (*19*). It is particularly important for helical macromolecules in which, even if the monomers are themselves optically inactive, the helical conformation renders the macromolecule optically active. That this property can vary with the molecular axes is demonstrated by considering the lateral mirror image of a helix. A right handed helix appears left handed in the mirror An end-on mirror image is a circle however. Hence, one would expect the observed θ to change if the solute macromolecules are caused to align as in an electric field. Tinoco (*20*) first verified this prediction using a 10% concentration of the polypeptide poly-(benzyl-L-glutamate) in a helix promoting solvent under high voltage continuous electric fields. Using the equation previously derived by himself and Hammerle (*21*) namely that

$$\Delta[\theta] = \frac{2}{3}\,([\theta_3] - [\theta_1])\,\Phi \tag{9}$$

with the non field value $[\theta]$ as

$$[\theta] = ([\theta_3] + 2\,[\theta_1])/3 \tag{10}$$

he obtained estimates for $[\theta_3]$ and $[\theta_1]$ at various wavelengths.

The method was demonstrated to have greater potential when measurements are made in the time domain using pulsed electric fields (*22*). In this case, not only are the amplitudes of the optical rotation changes recorded but also the relaxation times and hence the helical lengths can be evaluated.

For these measurements, the apparatus outlined in Fig. 2 can be used. The quarter-wave plate (Q) is removed, the analyser and polariser are crossed. The cell must allow the electric field to be concomittant with the light path if birefringence effects are to be avoided. In addition, one should work at a wavelength away from a dichroic absorption band. As the optical rotation is directional with respect to the propagation direction of the light beam (*i. e.* rotates specifically to the right or to the left for a given system) one cannot use the multireflecting cell. A cell of the type (*c*) of Fig. 6 is employed. This gives a long path length (and hence a larger observed θ) whilst maintaining the field strength through the solution. The electrodes are thin stannous oxide coatings on the appropriate inner faces of the quartz windows. After filling the cell, the procedure is as follows. Measure the rotation θ in the absence of any electric field by suitably offsetting P or A to restore the zero transmission condition. Make appropriate allowances and corrections for any solvent and cell window contributions. Upon application of a suitable pulsed electric field to the family of electrodes, a transient response will be detected by the photomultiplier. This is photographed. Such a response is shown in Fig. 9. The parameter τ can be obtained from the decay whilst the amplitude of the steady response can be recorded and measured as a function of the field strength and frequency and analysed in the usual manner.

Using the multiple cell described above and a low power He-Ne laser as light source, a solution of PBLG in ethylene dichloride of only 10^{-3} g ml^{-1} concentration was studied (*22*). In this particular case, it was not possible to generate fields of sufficiently high amplitude to approach the condition of complete orientation saturation as when Φ goes to unity. Values of $[\theta_3]$ and $[\theta_1]$ were therefore obtained

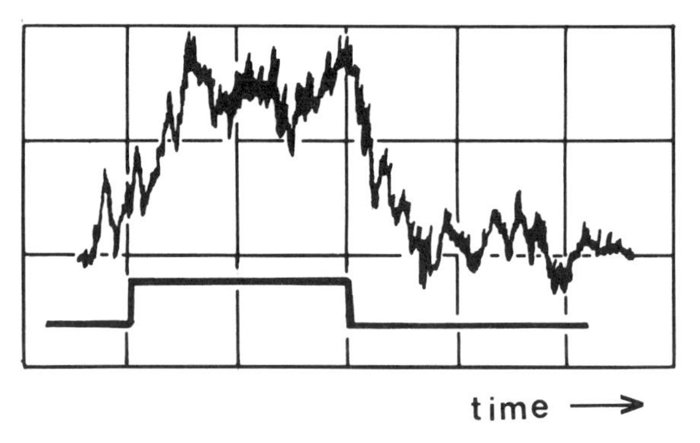

Fig. 9. Transient change in the optical rotation for poly-(benzyl-L-glutamate) in ethylene dichloride. Data for $c = 0.1\%$, at $\lambda = 633$ nm, with a field of 5 kV cm^{-1} amplitude and 1 ms duration. The maximum rotation change corresponds to 0.006 degrees. Data from Ref. (*22*)

by using the cell types Fig. 6a and c and making simultaneous measurements of the optical rotation and electric birefringence. From these Φ was found for a given field strength. Then $[\theta_3]$, $[\theta_1]$, μ_3, τ and hence the helix length were all found from the transient optical rotation data. The results agreed with values in the literature.

The advantages of the electric optical rotation method, beyond those common to the other electro-optic experiments, are as follows. Not only can one use θ as an empirical measure of changes in conformation of macromolecules, but one can simultaneously evaluate such gross parameters as the size and dipole moments of the molecules. In addition, evaluation of the specific directional parameters $[\theta_i]$ are useful in indicating the direction and origins of the relevant electronic transitions and in relating the molecular structure to model compounds. It should be mentioned however that the electrically induced changes in θ are small and far less easily detected to date than the other experiments described in this article.

V. Electric Light Scattering

Measurement of the light scattered by polymer solutions has become a standard method for determining the molecular weight and radius of gyration (S) of polymers. These molecules are generally such that their major dimension is of the order of the wavelength (λ) of visible light. Intramolecular interference is then encountered between the wavelets scattered from different regions of any molecule. Through such an interference pattern, the variation of the scattered intensity (I) with the angle of observation (θ) inherently contains information about the size and shape of the molecules. The basic scattering power of the particle irrespective of its geometry can be estimated from the intensity of the light beam scattered in the straight through or forward direction for which $\theta = 0$. Conventionally, measurements are made on dilute solutions at a variety of concentrations and angles of observation. From these the molecular weight of the polymer is obtained by analysing the zero angle scattering whilst the radius of gyration is evaluated from the angular variation of I. The experimental data are usually presented in the form of a Zimm plot. Details are given in a number of good reviews (*23–24*).

The intramolecular interference factor is strongly dependent on the orientation of the individual molecules (*25*). Hence, the scattered intensity and its angular dependence can be used to indicate statistically significant changes in the orientation of the molecules. Measurements of I and its changes (ΔI) upon application of electric fields to the solutions can thus be used to obtain information on the electrical parameters μ_i and α_i and on the relaxation time τ through experiments analogous to those described for the other electro-optic effects. Although restricted to molecules of sufficient size, the method has the great advantage of giving a wealth of molecular parameters when pre-field and in-field data are combined.

The apparatus is schematically presented in Fig. 10. It can be a suitably modified commercial scattering photometer in which the photodetector has a facility for recording both I and ΔI at various angles θ. A typical cell, in which the field can be applied perpendicular to the plane containing the incident and scattered beams, is shown in Fig. 6d. An alternative cell with electrodes of variable angle with respect to the direction of observation is also shown. Ref. (*26*) demonstrates how commercial photometers can be modified to give the additional sensitivity required for the electric field measurements.

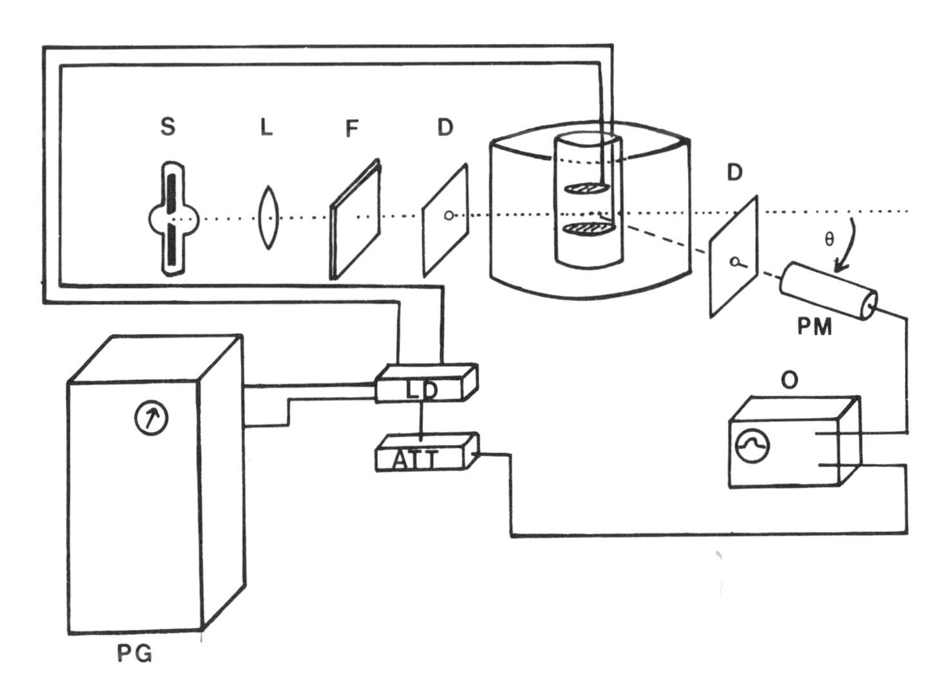

Fig. 10. Schematic of the apparatus used for electric field light scattering. The optical components are S – arc, L – lens, F – optical filter, D – diaphragms and PM – photomultiplier. The electronics are identical to those of Fig. 2. The scattering cell is immersed in a liquid-filled vat to minimise solution/cell/air infacial reflections

The theory for the scattering changes cannot be expressed as simply as for the foregoing methods. Equations have been published for rigid rods and discs and for flexible and 'frozen' coils which are either polar or exhibit a polarisability anisotropy. For the rigid molecules, at low degrees of orientation,

$$\Delta I/I = (1-3\cos^2\Omega)\,Q\left(\frac{(\mu_1^2-\mu_2^2)}{(1+\omega^2/4D^2)k^2T^2}+\frac{(\alpha_3-\alpha_1)}{kT}\right)E^2 \tag{11}$$

where Ω is the angle between the electric field vector and the scattering vector $\bar{s}$. This vector $\bar{s} = \bar{i}_\theta - \bar{i}_0$ where $\bar{i}_\theta$ and $\bar{i}_0$ are unit vectors in the observation and incident directions respectively. With the first cell described in Fig. 6d, Ω is always $\pi/2$. The factor Q is independent of the electrical properties of the molecules but does vary with the particle shape. For rods of length L and discs of diameter d, with $x = \frac{(2\pi)}{\lambda}\cdot\xi\sin\theta/_2$ where ξ represents L or d as appropriate,

$$Q_{\text{rod}} = \frac{1}{12}+\frac{1}{8}\frac{1}{P_0}\left(\frac{\sin 2x}{2x^3}-\frac{1}{x^2}\right) \text{and } Q_{\text{disc}} = -\frac{x^2}{180\,P_0} \tag{12}$$

Here P_0 is the particle scattering factor for the particular model in the absence of any electrical field. An advantage of electric light scattering is that the parameters preceding the brace of Eq. (11) can be evaluated from the scattering data in the absence of the field. One does not have to use the higher fields to obtain orientation saturation conditions. Hence, two consecutive pulsed scattering transients as indicated in Fig. 3 are all that are needed to fully characterise the nature and directions of the permanent and induced dipole moments. An example is given in Fig. 11 for an aqueous suspension of Laponite synthetic clay particles. The amplitudes of the transients indicate a *greater* ΔI at the higher frequency. This is equivalent to a dispersion curve which *increases* with frequency and indicates that, in the dc field, the permanent and induced contributions compete and are thus of opposite sign. In the present case, $\mu_3 > \mu_1$ and is predominantly pointing through the face of these disc-shaped particles whilst $\alpha_1 > \alpha_3$ and the greater polarisability is in a direction *along* the disc face. In this study the relative values of the electrical parameters were $(\mu_3^2-\mu_1^2)/(\alpha_1-\alpha_3) = 1.3 \times 10^{-21}$ J. The transient decay was equivalent to the sum of at least two exponential contributions [see Eq. (4)] with $D = 7.5$ and $3.3\ s^{-1}$. These corresponded to disc diameters of 740 nm and 980 nm respectively. The Zimm plot data gave $\bar{d} = 990$ nm which was in remarkably close agreement to that obtained from the transient decay curve. Furthermore, by incorporating the Zimm plot data for P_0 and d in Eqs. (11) and (12), discrete values of $(\alpha_1-\alpha_3) = 2.5 \times 10^{-29}$ Fm^2 and $(\overline{\mu_3^2})^{1/2} = 5.7 \times 10^{-25}$ Cm were obtained.

Equations for flexible chain molecules are very different from Eqs. (11) and (12). They have only been solved for the specific situations with $\Omega = 0$ and $\Omega = 90$. A full discussion of these is made elsewhere (*26*). Suffice it here to say that the changes ΔI under these two experimental conditions show a different dependence on the degree of polymerisation (N). Hence, electric field light scattering affords us a quick method of assessing whether a molecule is flexible or not. By measuring the ratio

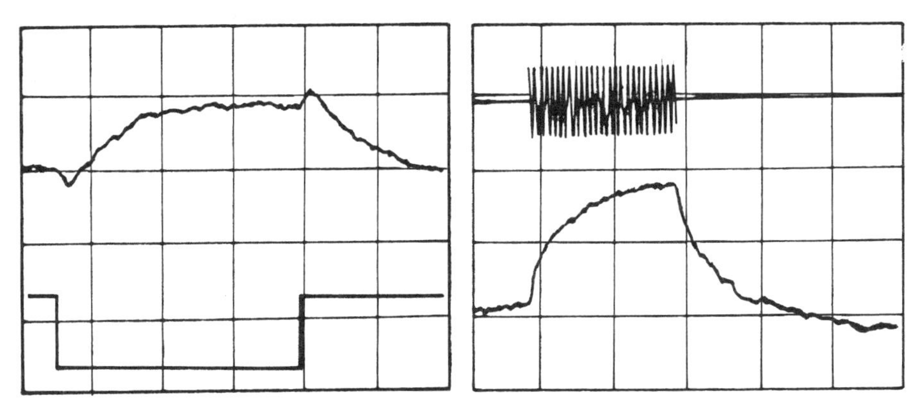

Fig. 11. Transient scattering responses for dc and ac pulsed fields. Data for a suspension of Laponite clay discs in fields of 80 V cm^{-1}. The dc field was of 70 ms duration, and the ac field of 100 ms duration with a frequency of 250 Hz. Data from Ref. (*27*)

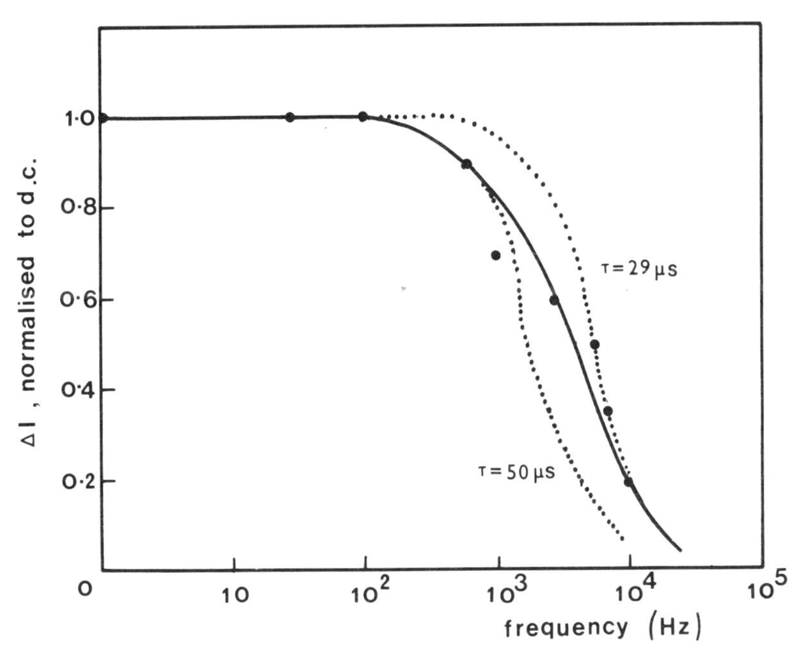

Fig. 12. Frequency dispersion curve of the scattered intensity changes for nitrobenzene in acetone. Data for $\theta = \Omega = 90°$, $\lambda = 436$ nm, $E = 710$ V cm^{-1}. At low frequency, the data fit a Debye dispersion with $\tau = 50\ \mu_s$ whilst at high frequency they agree with $\tau = 29\ \mu_s$. This is due to the sample polydispersity and partial flexibility. Data from Ref. (*29*)

$R = (\Delta I)_{\Omega=0}/(\Delta I)_{\Omega=90}$ a value of -2 is obtained with rigid molecules. For polar, flexible molecules this ratio is a complicated function of N and is far from -2.

This ratio was used in a study on an aqueous solution of the lefthanded helical form of polyproline II. At low field strength, ΔI varied with E^2. However, $(\Delta I)_{\Omega=90}$

was negligible when compared with $(\Delta I)_{\Omega=0}$ indicating that, although this polymer is helical, at high molecular weights the helix behaves in an electric field rather more like a flexible chain than a rigid cylinder (*28*).

A study has also been made on a sample of nitrocellulose having an average degree of nitration of 4.6 nitro groups per cellobiose unit (*29*). A Zimm plot was obtained for this sample in acetone and indicated a weight average molecular weight of 4.6×10^5 and a radius of gyration of 101 nm. In an electric field the molecule did not behave as a random coil using R as the criterion. Neither was the molecule completely rigid. The molecule reacted as if it had a strong permanent dipole moment directed along its molecular backbone. Any induced dipole was relatively insignificant (Fig. 12). It was estimated that the dipole moment was of the order of 5 debye units (16×10^{-30} Cm) per monomer unit. Such a result was in agreement with values obtained from dielectric measurements. In addition, a consideration of models of the cellubiose unit with this degree of nitration does indicate the likelihood of a resultant axial dipole moment. Hence, the method appears to give credible molecular parameters and indicate the semi-flexible nature of the polymers. A suitable theory for the electro-optic scattering from molecules with intermediate flexibility between the extremes of the rigid rod and the freely jointed chain is urgently required for the advancement of the method. The host of molecular parameters, the indication of rigidity and flexibility and even polydispersity (*30*) plus the speed and ease of the measurements make this an extremely promising characterisation tool.

VI. Conclusion

The imposition of order on macromolecules in dilute solution enables various axial components to be determined of those optical properties which exhibit anisotropy as a result of the molecular structure. These components of the tensor functions concerned may be used in certain favourable instances to indicate or confirm a proposed molecular structure.

Electric fields are of particular value as the means of inducing the order as they enable electric dipole moments and polarisabilities to be measured. These are themselves of value in indicating the molecular conformation of polymers and biopolymers.

The optical properties of polymer solutions are already widely used to characterise such solutions. The supplementary information on τ, μ and α is obtained from the same samples and under the same conditions as the data obtained from the existing experiments.

The methods are quick and relatively easy to perform; especially with molecules having a large electrical anisotropy (such as highly polar molecules). Macromolecules are ideally suited to this requirement.

Molecular flexibility can be estimated, but only to a limited extent. Because of the lack of suitable theories for the electro-optic behaviour of semi-flexible molecules, dynamic relaxation properties are generally studied. The field-free relaxation time τ can be analysed in terms of the persistence length concept using suitable equations. In addition, the scattering parameter R may be used.

The methods can be extended to any optical tensor property of the molecule. An example is electrically induced fluorescence (*31*). In addition, many liquid crystal display devices operate on electro-optic principles. Yet highly concentrated solutions of certain macromolecules form liquid crystalline phases. The electro-optic properties of solutions of ever-increasing polymer concentration therefore should be significant in the study of the nature and origins of liquid crystal properties.

Finally, with molecules having a high degree of flexibility or small size, the applied electric field needs to be of increased amplitude and, in the case of pulsed fields, of very short duration. Such pulses are extremely difficult to generate. Over the past three years a method has been developed in the author's laboratory in which the high intensity but short lived electric vector in a high powered switched laser beam has been used to induce order in macromolecular solutions. An example is given in Fig. 13. At the high frequency of the optical field, only the electronic processes can contribute to the electrical polarisability. The laser method extension should thus provide a means of both looking at faster orientation processes and isolating the electronic contribution to α.

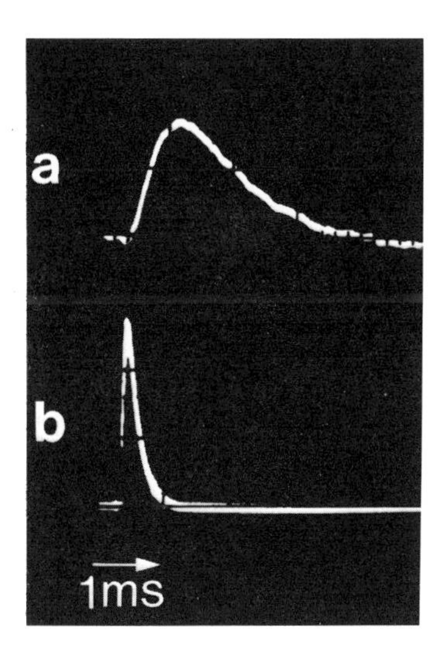

Fig. 13. Laser induced birefringence for a solution of tobacco mosaic virus in orinsens buffer at pH 7. Data for $c = 3 \times 10^{-3}$g ml^{-1}. The birefringence was detected using an Argon-ion laser at 488 nm wavelength. The inducing beam was from a YAG infra-red laser, at $\lambda = 1.06$ μm. The equivalent electric field was of approximately 5 kV cm^{-1} amplitude. The decay rate corresponded to $\tau = 1.0$ ms. Frame (a) is the photomultiplier response and (b) is the inducing laser pulse. The frames are not accurately synchronised in time. Reproduced from Ref. (*32*), by courtesy of 'Nature'

VII. References

1. Kerr, J.: A new relation between electricity and light: Dielectrified media birefringent. Phil. Mag. **50**, 337–348 (1875).
2. Yoshioka, K. and Watanabe, H.: Dielectric properties of proteins II Electric birefringence and dichroism: Physical principles and techniques of protein chemistry (Ed. S. Leach), p. 335–367. New York: Academic Press 1969.
3. O'Konski, C. T.: Kerr Effect in: Encyclopedia of polymer science and technology. New York: J. Wiley 9, 551–590 (1969).
4. Fredericq, E. and Houssier, C.: Electric dichroism and electric birefringence. London: Clarendon Press (1973).
5. Baily, E. D. and Jennings, B. R.: An apparatus for measurement of electrically induced birefringence, linear dichroism and optical rotation of macromolecular solutions and suspensions. J. Colloid and Interface Science **45**, 177–189 (1973).
6. Peterlin, A. and Stuart, H. A.: Doppelbrechung, insbesondere künstliche Doppelbrechung. Handbuch und Jahrbuch der Chemischen Physik 8, sec 1 b, 1–115. Leipzig: Becker & Erler (1943).
7. Broersma, S.: Rotary diffusion constant of a cylindrical particle, J. Chem. Phys. **32**, 1626–31 (1960).
8. Perrin, F.: Mouvement brownien d'un ellipsoide. I Dispersion diélectrique pour des molecules ellipsoidales. J. Phys. Rad. **5**, 497–511 (1934).
9. Berger, M. N.: Addition polymers of monofunctional isocyanates. J. Macromol. Sci., Revs. Macromol. Chem. **C9** (2), 269–303 (1973).
10. Jennings, B. R. and Brown, B. L.: The physical properties of polyisocyanates in solution. Europ. Polymer J. **7**, 805–826 (1971).
11. Yu, H., Bur, A. J., and Fetters, L. J.: Rodlike behaviour of poly (n-butyl) isocyanate from dielectric measurements. J. Chem. Phys. **44**, 2568–2576 (1966).
12. Hearst, J. E.: Rotary diffusion constants of stiff-chain macromolecules. J. Chem. Phys. **38**, 1062–1065 (1963).
13. Baily, E. D. and Jennings, B. R.: Simple apparatus for pulsed electric dichroism measurements. Applied Optics **11**, 527–532 (1972).
14. Jennings, B. R. and Baily, E. D.: Transient electric dichroism of macromolecules using a simple spectrophotometer. Nature **233**, 162–163 (1971).
15. Foweraker, A. R. and Jennings, B. R.: Cells for simple longitudinal electric dichroism measurements. Laboratory Practice.
16. Robertson, J. M.: An X-ray study of the structure of the phthalocyanines. Part I. The metal-free, nickel, copper and platinum compounds. J. Chem. Soc. Part I, **1935**, 615–621.
17. Foweraker, A. R. and Jennings, B. R.: Orientation of the electronic transitions in crystalline copper phthalocyanine by means of electric dichroism. Spectrochim. Acta **31 A**, 1075–1083 (1975).
18. Fresnel, A.: La Double Réfraction. Oeuvres **2**, 479 (1868).
19. Partington, J. R.: An advanced treatise on physical chemistry, Vol. IV, p. 303. London: Longmans 1953.
20. Tinoco, I.: The optical rotation of oriented helices. I Electrical orientation of poly-γ-benzyl-L glutamate in ethylene dichloride. J. Amer. Chem. Soc. **81**, 1540–1544 (1959).
21. Tinoco, I. and Hammerle, W. G.: The influence of an external electric field on the optical activity of fluids. J. Phys. Chem. **68**, 1619–1623 (1956).
22. Jennings, B. R. and Baily, E. D.: Transient electric optical rotation for macromolecular characterisation. J. Polymer Sci. Symp. **42**, 1121–1130 (1973).
23. Stacey, K. A.: Light scattering in physical chemistry. London: Butterworths 1956.
24. Huglin, M. B. (editor): Light scattering from polymer solutions. London: Academic Press 1972.
25. Wippler, C.: Diffusion de la lumière par les solutions macromoléculaires. J. Chim. Phys. **53**, 316–351 (1956).
26. Jennings, B. R.: Electric field light scattering, Chapter 13 of Ref. (*24*).

27. Schweitzer, J. F. and Jennings, B. R.: Transient scattering changes induced by pulsed sinusoidal electric fields. J. Phys. D. (Appl. Phys.) **5**, 297–309 (1972).
28. Jennings, B. R.: Structural information from the light scattered by solutions subjected to electric fields. Brit. Polymer J. **1**, 70–75 (1969).
29. Jennings, B. R. and Schweitzer, J. F.: Electro-optic scattering from nitrocellulose solutions. Europ. Polymer J. **10**, 459–464 (1974).
30. Stoylov, S. P. and Sokerov. S.: Transient electric light scattering. II Determination of distribution curves for solutions of polydisperse rods. J. Colloid Interface Sci. **27**, 542–546 (1968).
31. Weill, G. and Hornick, C.: Polarisation of fluorescence of an orientated solution of rodlike particles bearing a fluorescent label. Biopolymers **10**, 2029–2037 (1971).
32. Jennings, B. R. and Coles, H. J.: Laser-induced orientation in macromolecular suspensions. Nature **252**, 33–34 (1974).

Received February 3, 1976

Ultrasonic Degradation of Polymers in Solution

Arno Max Basedow and Klaus Heinrich Ebert
Institut für Angewandte Physikalische Chemie, Universität Heidelberg, D-6900 Heidelberg

Table of Contents

1. Introduction

Ultrasonic degradation of macromolecules in solution is a special case of a disintegration reaction of the macromolecules induced by a complicated action in which hydrodynamic forces are of primary importance. The particular interest of ultrasonic degradation is the fact that, contrary to all chemical or thermal decomposition reactions, ultrasonic depolymerization is a *nonrandom* process that produces fragments of definite molecular size. If the theoretical principles of the mechanism of the process are known, the production of polymer fractions having definite molecular-weight distributions seems possible. By the action of ultrasound on polymer solutions large molecules are more rapidly degraded than smaller ones; this process therefore permits the removal of large molecular-weight tails commonly present in technically produced or fractionated polymers, which is otherwise difficult to achieve. This is one of the most interesting applications of ultrasonic degradation in polymer chemistry, since large molecular weight components usually reduce the mechanical properties and for this reason the general quality of the polymer. The existence of a limiting molecular weight, below which degradation by ultrasound does not take place, has the additional effect that initially broad molecular-weight distributions become more narrow during irradiation.

Ultrasonic degradation is much more effective than the degradation of polymers in solution by hydrodynamic shear, and since the laws governing these two processes are similar in many aspects, ultrasonic degradation may be used in many experiments on shear degradation if the effects of the latter are too small for experimental detection. The most important application of ultrasound in this respect is the investigation of the stability of polymer additives widely used in multigrade lubricating oils. Ultrasound also has a strong liquefying effect on natural and synthetic gels and is therefore used extensively in technological operations. In biochemistry ultrasound is commonly used for the disruption of cells and living tissue in order to extract soluble constituents more efficiently. Since most proteins, enzymes and nucleic acids are easily denatured by ultrasound, the knowledge of the kinetics and the mechanism of this degradation reaction is essential for its use in biologic sciences. It is the aim of this review to provide a detailed discussion of the kinetics and the mechanism of ultrasonic degradation of polymers in solution, in the light of the experimental results and theoretical knowledge available today.

2. Effects of High Intensity Ultrasonics in Liquids

2.1. Elementary Physics of Ultrasonic Waves

In ultrasonics we are concerned almost exclusively with sinusoidal motion and, in the majority of applications in liquids, with longitudinal waves only. Though transverse waves are also propagated in liquids, they need not normally be regarded because their attenuation with distance is extremely high. The general laws of acoustics can also be applied to ultrasonic waves; an additional advantage of ultrasound is that sharp focusing is possible and that high intensities can be generated.

The velocity of sound c in liquids, as evaluated from the theory of wavemotion, depends on the adiabatic compressibility β and the density ρ of the medium; it is given by the equation

$$c = \lambda \nu = \sqrt{\frac{1}{\beta\rho}} \quad (1)$$

where λ and ν denote the sound wavelength and frequency respectively. The velocity of ultrasound in water is approximately 1500 m/sec; the wavelengths corresponding to frequencies from 20 kc/sec to 1000 kc/sec fall within the range of 7.6 cm to 0.15 cm. The frequencies between 20 kc/sec and 1000 kc/sec are commonly used in the investigations of ultrasonic chemical processes.

A sound wave propagated in a medium carries a certain amount of energy with it, which is generally characterized by the *intensity* of the sound J (watts/m^2), and which is defined as the energy transmitted in one second through a unit area perpendicular to the direction of propagation. For a plane progressive wave in a liquid, the peak values of the alternating components of pressure P_0, displacement A_0, velocity U_0 and acceleration B_0 are given by:

$$P_0 = \sqrt{2\rho c J} \quad (2)$$

$$A_0 = \frac{1}{2\pi\nu} \sqrt{\frac{2J}{\rho c}} \quad (3)$$

$$U_0 = \sqrt{\frac{2J}{\rho c}} \quad (4)$$

$$B_0 = 2\pi\nu \sqrt{\frac{2J}{\rho c}} \quad (5)$$

The quantity ρc is called the *specific acoustic impedance* of the medium; it denotes the ratio of the acoustic pressure to the velocity of the vibrating particle. In acoustics the relative sound intensity level is usually expressed by means of a logarithmic quantity k, measured in decibels (db). If sound of intensity J_1 is compared with sound of intensity J_2, k is given by:

$$k = 10 \log \frac{J_1}{J_2} \quad (6)$$

For reference measurements, the sound intensity J_2 is taken close to the threshold of audibility of the human ear, which is approximately 10^{-16} watts/cm^2. The average intensity of ultrasonic waves used in laboratory investigations of liquids, which is in the order of 10 watts/cm^2, is therefore many orders of magnitude greater than the sound intensity of the most powerful loudspeakers. As an example let us consider the irradiation of water at an intensity of 10 watts/cm^2 and a frequency of 500 kc/sec (ρ = 1 g/cm^3, c = 1.48x10^5 cm/sec). From the above equations we obtain: P_0 = 5.4 bar, A_0 = 1.2x10^{-5} cm, U_0 = 36.8 cm/sec and B_0 = 1.1x10^8 cm/sec^2. These values show that the dynamic acoustic pressure varies 500000 times per second between +5.4 bar and −5.4 bar. The primary acoustic pressure, the particle amplitude and the velocity have relatively low values, so that exceptional effects are not to be expected from them. The acceleration of the particles of the liquid, however, can be very large; in the case considered the acceleration is 100000 times greater than the action of gravity. The rapid variation of the acoustic pressure causes the formation of small cavities in the liquid, which, as will be shown later, are responsible for most effects of high-intensity ultrasonics in liquids.

When an acoustic wave travels through matter its intensity decreases exponentially with distance, the acoustic pressure being given by the equation:

$$P_x = P_0 \cdot e^{-\alpha x} \cdot e^{i(2\pi \nu t - 2\pi x/\lambda)} \tag{7}$$

where P_x is the pressure at a distance x from the source, α is the absorption coefficient; the second exponential term refers to the sinusoidal wavemotion. The energy lost from the radiation appears as heat. In liquids absorption of sound is a complex process, which is due mainly to viscosity, thermal conductivity and relaxation processes. Since the two latter phenomena are often very small in pure liquids, compared to the first, a simplified equation, considering the internal friction forces only, may be employed in most cases. Based on Stoke's theory, the following expression for the *absorption coefficient,* which considers viscosity only, can be evaluated:

$$\alpha = \frac{8\pi^2 \nu^2 \eta}{3\rho\, c^3} \tag{8}$$

where η is the viscosity of the liquid. The reciprocal value of α is called the *depth of penetration.* In polymer solutions, however, the calculation of absorption coefficients is more complicated because of the existence of various relaxation times, corresponding to a whole series of molecular processes. For more details on this subject see Flügge (*19*), Bergmann (*5*), Gooberman (*25*), Mason (*50*), Blandamer (*8*),and Nosov (*67*).

2.2. Generation of Ultrasound in Liquids

In most experimental techniques, ultrasonic vibrations are introduced into the liquid from a solid. This is possible either by direct contact of the liquid with the ultrasonic transducer or indirectly by means of a coupling member, the so-called ultrasonic horn. The use of horns avoids the contact of the liquid that is being irradiated with the transducer. If necessary, the intensity can be increased over small volumes by characteristic geometrical forms of the horn. Usually cylindrical, conical, catenoidal or exponential horns are used (*5*). In the latter case the vibration amplitude increases along the bar inversely with the cone diameter. The dimensions of ultrasonic horns must be such that the natural resonance frequency of the vibrator is not appreciably affected. The vibrator or transducer is invariably excited electrically, using either piezoelectric or magnetostrictive materials. Transducers supplied by mechanical energy, *e.g.*, the various types of ultrasonic whistles, are widely employed in ultrasonic processing of solutions and in emulsification. They are, however, rarely used in laboratory practice.

The most important piezoelectric materials used for the generation of high-intensity ultrasonics are quartz and a variety of electrically polarized ceramic substances, such as barium titanate and lead zirconate titanate. Using ceramics, more than 90% of the electrical energy applied to the plate can be transformed into ultrasonic energy. Ceramics are mainly used nowadays as sources of ultrasound, because of their good mechanical properties, especially their high stability even when the vibrations are of considerable amplitude. Piezoelectric materials may be used either directly immersed in a nonconducting liquid or coupled to a resonant horn; in the latter case intensities up to 500 watts/cm^2 are attainable. Among the group of magnetostrictive materials used as ultrasonic generators, pure nickel and iron–cobalt alloys are mainly used. The efficiency of these transducers depends largely on the frequency; values of about 20% for high frequencies and 70% for low frequencies are good averages. Magnetostrictive transducers are always used in connection with a coupling horn; intensities up to 100 watts/cm^2 are obtainable.

For most practical applications of ultrasound in liquids it is desired to have the ultrasound irradiated from one end of the transducer only. To achieve this, the other end of the transducer is placed in a medium with a specific acoustic impedance as low as possible. The ideal medium would be vacuum, but if the transducer is designated for the irradiation of liquids, which generally have a large acoustic impedance, the nonradiating end may be left in contact with air.

Piezoelectric transducers must be fitted with electrodes, which supply the necessary alternating electrical field. These electrodes must not impede the irradiation of the ultrasonic waves; therefore thin gold or silver films are usually deposited on the transducer surface by vacuum evaporation. A crystal prepared in this way can be used directly for the generation of ultrasound in a nonconducting liquid, *e.g.*, oil. A typical experimental set-up of this type used for the irradiation of liquids is shown in Fig. 1. The main disadvantage of this arrangement is the difficulty in measuring

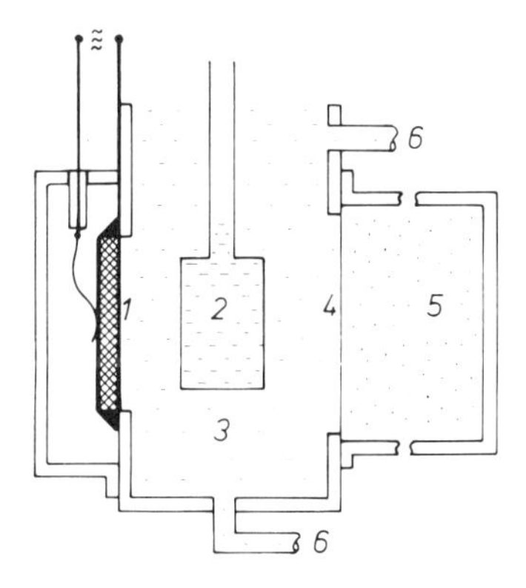

Fig. 1. Experimental degradation apparatus with simple crystal holder.
1: Crystal or ceramic plate *2:* Glass cell with polymer sample *3:* Paraffin or transformer oil *4:* Membrane *5:* Absorber *6:* Connections to thermostat

exactly the intensity of the ultrasound that is actually acting on the liquid, because of the inevitable reflections at the walls of the reaction vessel. If the reaction vessel, however, is large enough, a monitoring device for ultrasonic intensity can be inserted directly into the liquid during irradiation. The temperature control of the system is very good. In order to assure homogeneous irradiation, the distance between the reaction vessel and the ultrasonic vibrator must be larger than one wavelength of the ultrasound. For ultrasonic generators provided with a horn, an arrangement as shown in Fig. 2 is generally applied. For high-intensity applications the main

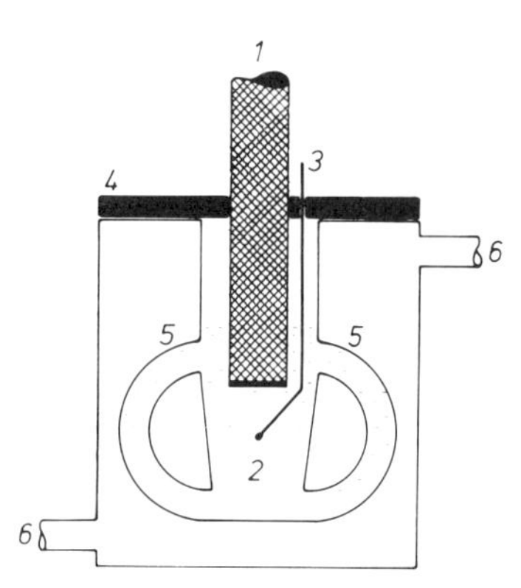

Fig. 2. Cooling cell for degradation experiments. *1:* Ultrasonic vibrator *2:* Polymer solution *3:* Thermocouple *4:* Rubber seal *5:* Side-arms for circulation of the solution *6:* Connections to thermostat

disadvantage in this case is the poor temperature control. A special shape of the reaction vessel, however, can improve temperature stability considerably. In the vessel shown in Fig. 2, the ultrasonic radiation pressure emitted from the horn forces the liquid to flow through the side arms, thus effecting good mixing and an efficient means of heat exchange. In all practical cases the wavelength of the ultrasound is larger than the diameter of the vibrating tip of the horn, so that the ultrasonic field is almost homogeneously irradiated. The ultrasonic intensity can be easily obtained from the electrical output of the high-frequency generator and the efficiency of the transducer system.

2.3. Cavitation

If a liquid is irradiated at low ultrasonic intensities no observable effects occur apart from a slight heating due to the absorption of sound. On increasing ultrasonic intensity, so that the acoustic pressure amplitude reaches values of the order of 1 bar small foggy bubbles appear in the liquid. At this point usually a hissing or tearing sound can be heard. The formation of these bubbles in the liquid and their subsequent collapse is called *cavitation.* Cavitation is a three-step process consisting of nucleation, growth and collapse of these gas or vapor-filled bubbles in the liquid. A cavitation bubble is formed by the pressure variation due to the ultrasonic wave, when the temporary reduction of pressure in the liquid falls below the threshold of tensile strength of the liquid. Since the tensile strength of pure homogeneous liquids is of the order of hundreds of bars, but the acoustic pressure amplitude is of the order of bars only, cavitation should not normally occur in liquids. The presence of microparticles, dissolved gases or other *cavitation nuclei* considerably reduces the liquid strength. Under the action of tensile stress, the liquid surface can then recede from the nucleus giving rise to a macroscopic strain and consequently rupture of the liquid. Cavitation nuclei may consist of small gas bubbles suspended in the liquid or bubbles stabilized in cracks of suspended solid particles or at liquid–solid interfaces or they may arise when high-energy elementary particles pass through the liquid or electro-chemical processes take place. Since surface tension and hydrostatic pressure cause very small gas bubbles to dissolve in the liquid, but larger ones to coalesce, expand and rise to the surface, it seems very unlikely that the liquid should contain stable gaseous nuclei of a size large enough to permit cavitation. In fact such nuclei exist probably as a result of stabilization by thin surface films of impurities. For water that has been standing for several hours at room temperature, a free nucleus would have to have a radius of 5×10^{-4} cm (20) or less, in order to remain stable for a reasonable period of time. Nuclei of such a radius will persist indefinitely in the liquid, if they are stabilized in some manner against diffusion.

The onset of cavitation in an irradiated liquid is a highly variable and complicated phenomenon, which is not yet fully understood. Since the rate at which the small bubbles coalesce and the cavity grows depends on the difference between the acoustic pressure amplitude and the tensile strength of the liquid, ultrasonic energy must be applied for a certain time before cavitation takes place. Once a liquid has cavitated for some time, it usually cavitates at a lower ultrasonic intensity than that required to start cavitation. When high-intensity ultrasonic waves are propagated in a liquid, cavitation bubbles arise at the sites of rarefaction, *i.e.*, at the sites of negative pressure. The cavity formed expands, and then is filled with vapor of the surrounding liquid or with gases that are normally dissolved in the liquid. During one of the next compression phases the cavity collapses fairly rapidly.

The motion of a cavitation bubble is often complicated by sudden changes in the shape of the bubble, which does not always remain spherical. According to their motion, cavitation bubbles may be classified into *transient cavities,* which are cavities that contract from some maximum size, whereas the inward speed increases until the rapid rise of pressure within the cavity stops the inward motion, and *stable cavities,* which are cavities that oscillate nonlinearly about the equilibrium radius during many ultrasonic periods. Mundry and Güth (*30*) and Schmid (*82*) investigated the formation, growth and collapse of cavitation bubbles using high-speed cinematography. They showed that during the expanding period the bubbles remain relatively spherical; the same applies to the initial phase of the contraction, but in the later states of collapse, cavitation bubbles attain very irregular shapes. They may even disrupt into a large number of very small bubbles, which under the action of sound waves, can act as nuclei for the formation of new cavitation bubbles. The mathematical treatment of the formation and collapse of cavities in an ultrasonic field is complicated because of the numerous factors to be considered, such as frequency and amplitude of the exciting ultrasonic wave, compressibility of the liquid, thermal conductivity, evaporation and diffusion processes, surface tension, variation of pressure with time and nature of the nuclei present. The general problem in acoustic cavitation is to formulate the equation of motion of a cavity under the influence of ultrasonic waves. A detailed analysis of this problem has been given by Flynn (*20*). The effects of the initial wall velocity and the physical properties of the system, such as internal pressure of the cavity, density, viscosity and surface tension of the liquid, on the growth of cavitation bubbles have been studied by analyzing the equations of motion on an analog computer by Marique and Houghton (*48*).

Rayleigh (*80*) as early as 1917, developed an equation of motion for the collapse of a transient spherical cavity, assuming that both the pressure at infinity P_0 and the pressure within the cavity P_c were constant. Equating the kinetic energy of the liquid and the work done during the contraction, he calculated the velocity U of the interface of a cavity with radius R:

$$U^2 = \frac{2}{3} \cdot \frac{(P_0 - P_c)}{\rho} \left(\frac{R_0^3}{R^3} - 1 \right) \tag{9}$$

where ρ is the density of the incompressible fluid and R_0 is the radius at which the bubble starts to collapse. Eq. (9) shows that the velocity of the wall increases considerably as the cavity becomes smaller. If it is further assumed that the interface of a collapsing cavity strikes an absolutely rigid sphere of radius R, the instantaneous pressure P_* developed in the liquid is given by:

$$\frac{P_*^2 \cdot \beta}{2} = \frac{1}{2} \rho U^2 = \frac{(P_0 - P_c)}{3} \left(\frac{R_0^3}{R^3} - 1 \right) \tag{10}$$

where β is the compressibility of the liquid. As an example for water and taking that: $P_0 = 1$ bar, $P_c = 0$, $R = 1/20\ R_0$ and $\beta = 5 \times 10^{-5}$ bar^{-1}, an instantaneous pressure of 10^4 bars is calculated, showing that the pressure thus developed is very high.

The classical theory of the growth and collapse of a cavitation bubble in a liquid was developed by Noltingk and Neppiras (*64, 65*). Assuming that the liquid is incompressible and the gas content of the cavitation bubble remains constant, an equation has been developed describing the motion of a spherical cavity. The energy balance requires that the kinetic energy of the liquid is equated to the algebraic sum of the work done by the surface tension, the gas pressure and the liquid pressure at infinity, including the hydrostatic pressure on the liquid and the sinusoidal acoustic pressure variation due to the ultrasonic waves. The equation thus obtained, which shall not be reproduced here, is mathematically insoluble, but several special numerical solutions have been obtained on an analog computer. The calculated isothermal variation of radius with time for a gas-filled cavitation bubble is given in Fig. 3. The shapes of the calculated curves are very similar to the experimental time-radius curves obtained by Güth and Mundry (*30*), which are represented in Fig. 4. As an approximation, neglecting surface tension and assuming the internal pressure of the bubble to be zero, Okuyama and Hirose (*71*) evaluated an analytical solution of the equation of motion of a cavitation bubble under the action of rectangular ultrasonic waves. Typical time-radius curves calculated by this method are shown in Fig. 5. These curves are also similar in shape to those in Fig. 3, showing that the ultrasonic waveform has no pronounced influence on the motion of a cavity. In all cases it can

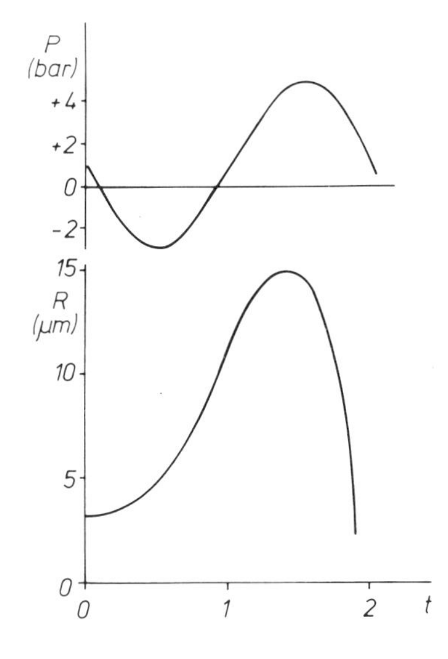

Fig. 3. Calculated variation of radius (R) with time (t) of a isothermal gas-filled cavitation bubble in water. P: acoustic pressure; ultrasonic frequency ≈ 500 kc/sec [Ref. (65)]

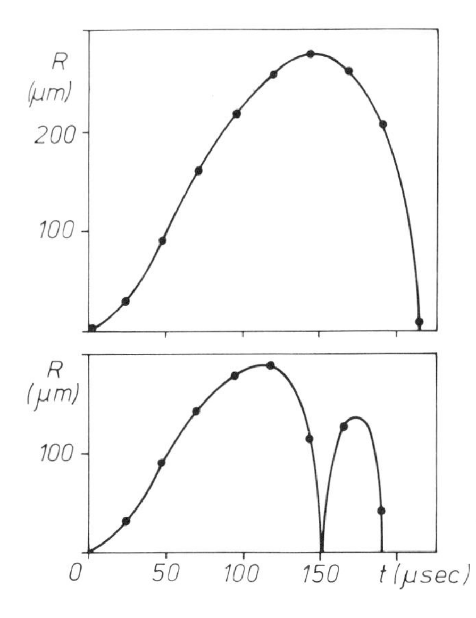

Fig. 4. Plot of mean radius (R) of a cavitation bubble in water against time (t).
Top: collapsing bubble *Bottom:* pulsating bubble; ultrasonic frequency 2.5 kc/sec [Ref. *(30)*]

be seen that the total collapse time is only a small fraction of the ultrasonic vibration period.

From the theoretical considerations of Noltingk and Neppiras *(64, 65)* many conclusions can be drawn. The most important result is that the maximum radius of the cavitation bubble is inversely proportional to the frequency of the ultrasound within a large range. Once the maximum bubble size has been reached, collapse follows in a manner determined mainly by the value of the maximum bubble radius. Therefore, cavitation effects diminish with increasing frequency. Noltingk and Neppiras also derived an expression for the pressure distribution in the liquid surrounding a cavity filled with gas and which collapses adiabatically. The results are depicted in

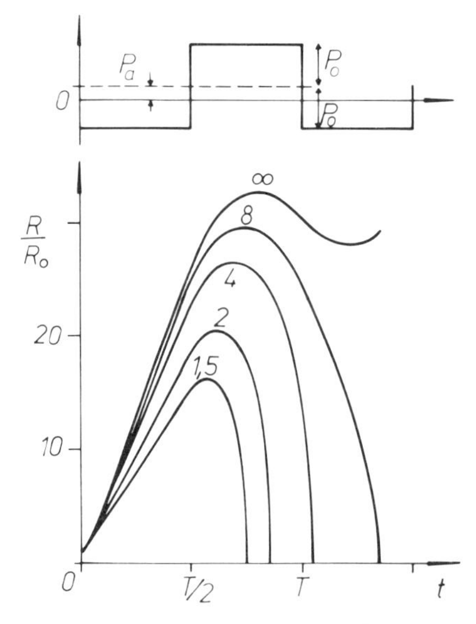

Fig. 5. Calculated variation of radius (R/R_0) of a void cavitation bubble with time (t) under rectangular ultrasonic waves. R_0: initial radius of the bubble T: period of ultrasonic vibration P_a: static pressure exerted on liquid P_0: acoustic pressure amplitude. Parameter of curves is P_0/P_a; for $P_a \to 0$ bubble starts to oscillate [Ref. (*71*)]

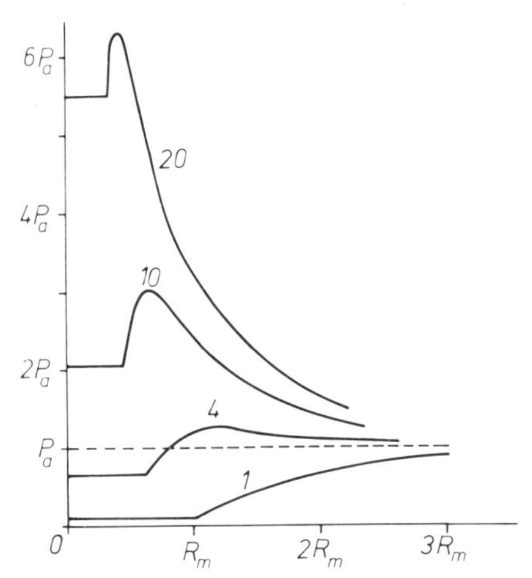

Fig. 6. Calculated pressure distribution in liquid surrounding gas-filled collapsing cavitation bubble. P_a: static pressure exerted on liquid R_m: maximum bubble radius. Parameter of the curves is $(R_m/R)^3$. Curves show that for relatively small changes in ratio R_m/R, maximum pressure P_m developed can be enormous [Ref. (*65*)]

Fig. 6, which shows that the amplitude of the pressure wave in a liquid following a collapsing cavitation bubble rises steeply as the bubble radius diminishes. The pressure can attain enormously high values, but their duration is only a small fraction of the total collapse time and their effects are concentrated over a very small volume of the liquid near the bubble surface only. The variation of this maximum liquid pressure with ultrasonic frequency is shown in Fig. 7. From this it follows that, if intense cavitation is desired, the ultrasonic frequency should be as low as possible.

Furthermore, Noltingk and Neppiras (*65*) and Jellinek and Brett (*42*) derived an equation for the boundary velocity of an adiabatically collapsing gas-filled cavity. This velocity depends markedly on the ratio of specific heats of the gas and its pressure inside the cavity at its maximum radius. The results of these calculations are given in Fig. 8. The position of the maximum of the velocity curve occurs always

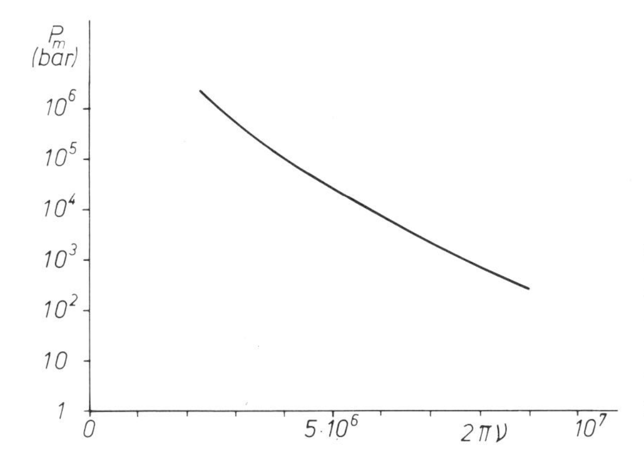

Fig. 7. Calculated maximum liquid pressure (P_m) during collapse of adiabatic gas-filled cavity as a function of frequency (ν) of ultrasound. Acoustic pressure amplitude = 4 bar [Ref. (*65*)]

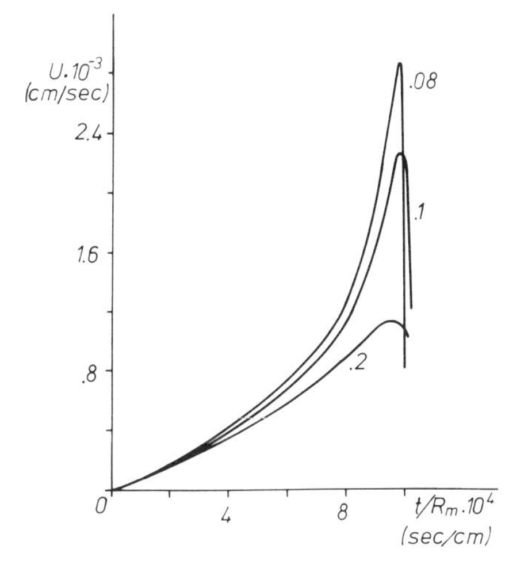

Fig. 8. Calculated wall velocity (U) of adiabatically collapsing gas-filled cavity in water at atmospheric pressure as a function of time (t). Parameter of curves is gas pressure (in bars) inside bubble; R_m: maximum radius of cavity [Ref. (*45*)]

at the same t/R_m value and is independent of the pressure inside the cavity. Once the numerical value of the velocity of the wall of a cavitation bubble is known, it is easy to calculate the velocity at any distance from the bubble by applying the equation of continuity. It is thus possible to compute the frictional force that acts on a polymer molecule of a given size present in the liquid near the collapsing cavity. It will be shown in Section 6.5 that this force may be large enough to break macromolecules, if certain additional conditions are fulfilled. Another interesting result from the theory of Noltingk and Neppiras is that the occurrence of cavitation is restricted to a definite range of variation of the following parameters: ultrasonic frequency, initial radius of the bubble nucleus, hydrostatic pressure, and acoustic pressure amplitudes. If the radius of a bubble is below a certain threshold, cavitation cannot occur because surface tension forces prevent the bubble from growing. According to Okuyama and Hirose (*70*) this is the case if the nucleus has a radius less than 1.7×10^{-5} cm. If the ultrasonic frequency is below the resonance frequency of a bubble, the expansion period during the half-cycle of rarefaction is followed by the rapid collapse in the compression half-cycle. However, if the ultrasonic frequency is greater than the resonance frequency of the bubble, the latter will not have reached its maximum radius before the ensuing negative ultrasonic pressure will cause it to expand again. The collapse of the cavity is avoided and a complex forced oscillation of the cavity with a comparatively small amplitude will occur. Cavitation is also impossible if the acoustic pressure amplitude is less than the hydrostatic pressure exerted on the liquid. Finally, cavitation is suppressed if the hydrostatic pressure is so small that appreciable quantities of liquid can evaporate into the bubble and thus exert a cushioning effect on the bubble walls during contraction. A more detailed discussion of the domains of cavitation is given by Okuyama and Hirose (*70*); a recent review on cavitation phenomena has been presented by Chendke and Fogler (*14*).

Many investigators have demonstrated that the real motion of a cavity in a liquid is highly damped due to the loss of energy, caused by heat conduction, sound radiation and viscous flow. These effects are more pronounced when the motion of the cavity begins to be transient. The most important of these effects on cavitation dynamics is thermal conductivity. The behavior of the cavity content concerning this latter effect is rather involved: at low and very high frequencies isothermal conditions are generally accepted, whereas at moderate and high frequencies the motion of cavities is almost adiabatic. Flynn (*20*) showed theoretically that the velocity of collapse of cavity where heat conduction takes place is always greater than that of a similar but adiabatic cavity. This is of course also the same for the kinetic energy transmitted to the liquid surrounding the collapsing cavity. This and the lower temperatures inside the cavity, are the main reasons why low-frequency ultrasound is more recommendable for degradation studies of polymers than high-frequency ultrasound.

If a small cavity is at rest in a liquid, surface tension and hydrostatic pressure will quickly cause it to dissolve. The alterations of pressure in an ultrasonic field, however, cause small bubbles to coalesce and the asymmetry and instability of its surface favors vapor or dissolved gas to diffuse from the liquid into the bubble. As a result, those bubbles may grow and finally reach a definite size at which violent oscillations occur produced by the ultrasonic waves. The fundamental frequency of the oscillations of such a bubble is given by the equation of Minnaert (*53*):

$$\nu = \frac{1}{2\pi R}\sqrt{\frac{3\gamma P}{\rho}} \tag{11}$$

where R is the radius of the bubble, γ the ratio of specific heats C_P/C_v of the gas inside the bubble, P the hydrostatic pressure and ρ the density of the liquid. The amplitude and energy of vibration of the gas bubble in an irradiated liquid is maximum in the case when its fundamental frequency is identical with the ultrasonic frequency. In this case of resonance the pressure in the neighborhood of the bubble can exceed the hydrostatic pressure by many orders of magnitude. As will be shown later, the large velocity gradients that are generated by these vibrations are capable of causing the rupture of macromolecules.

From the observations above can be understood that acoustic cavitation is an effective process for concentrating mechanical energy. Because the mechanical energy is accumulated into the very small volume of the cavities, many interesting phenomena, such as emulsification of immiscible liquids, erosion of solids, luminescence in liquids, initiation of chemical reactions and the disintegration of macromolecules are caused by cavitation. The absolute value of the total energy liberated during the collapse of a cavitation bubble is very small, but the energy density generated as a result of the converging motion of the bubble is enormous because of the focusing effect of the radial flow of the liquid. Cavitation therefore transforms the relatively low-energy density of a sound field into the high-energy density in the interior and at the boundaries of a collapsing bubble. From the practical point of view, cavitation provides the most effective source of mechanical energy capable of causing the specific degradation of macromolecules.

2.4. Shock Waves Produced by Cavitation

As shown in the last section the inward motion of the boundary of a collapsing cavity is suddenly arrested by the rising pressure of the gas and vapor in it, resulting in the development of large pressures in the liquid surrounding the cavity. According to the theory of Noltingk and Neppiras (*65*) the pressure variation in the liquid due to the adiabatic collapse of a gas-filled cavity is given in Fig. 6. This pressure wave will be radiated into the liquid as a shock wave. Initially at least, the form of this shock wave consists of a rather slow pressure rise in the leading edge, followed by a steep pressure fall in the tailing edge. This is the reverse of the shock wave normally generated due to the dispersion of the velocity of ultrasound. Waveforms similar to a reverse sawtooth have been also found experimentally for the shock waves formed by electric spark generated cavities, as well as for shock waves generated by violently vibrating metal surfaces (*29, 30*).

By applying the equation of Rayleigh, Güth (*29*) calculated the pressure distribution in an incompressible liquid during the collapse of a gas-filled cavity. The variation of pressure with the distance from the center of the cavity is given in Fig. 9 for different radii of the collapsing cavitation bubble. The shape of the curves is very similar to those obtained by Noltingk and Neppiras for gas-filled cavities (Fig. 6) in an incompressible liquid. At the moment of complete collapse, the maximum pressure would in this case be at infinity. This will, however, not occur in practice, since all liquids have a definite compressibility and some gas or vapor is always present in the bubble, so that finite values of pressure will be obtained. Güth also calculated the maximum pressure of the gas inside the bubble at its minimum radius, due to adiabatic compression, as a function of the gas content of the bubble. This gas content can be expressed as a function of the initial pressure of the gas inside the bubble at the starting moment of collapse. The maximum pressure attained, which is equal to the maximum value of the dynamic pressure wave radiated into the liquid, is shown in Fig. 10 as a function of the relative gas content of the bubble.

The propagation of the pressure wave thus generated is only possible if the liquid is assumed to have a finite compressibility. The exact mathematical treatment of this problem is very complicated and has not yet been attempted; but according to Güth the pressure variation is in principle similar to that in Fig. 9, with the only difference that the curves are not as steep as in the ideal case. In reality,

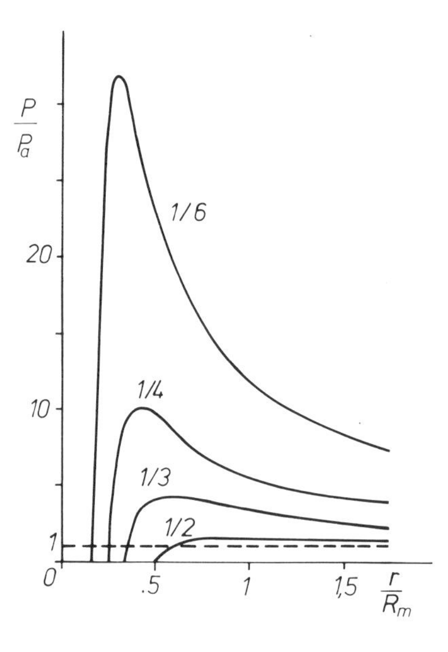

Fig. 9. Calculated pressure variation (P/P_a) in incompressible liquid during collapse of adiabatic gas-filled cavity as function of distance (r) from center of the cavity. P_a: static pressure exerted on the liquid R_m: maximum radius of the cavity R: radius of the cavity. Parameter of the curves is ratio R/R_m [Ref. (*29*)]

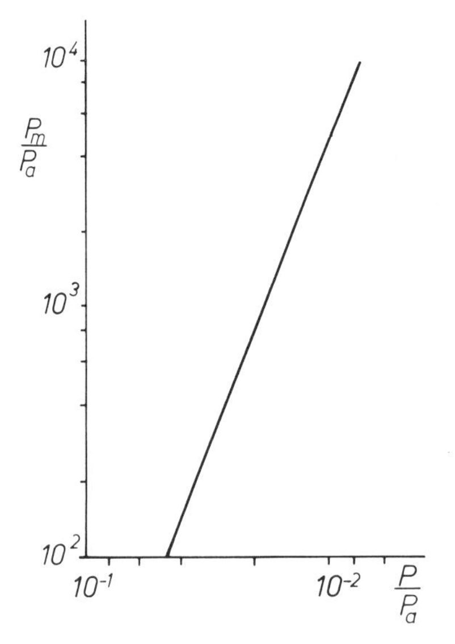

Fig. 10. Calculated maximum pressure P_m/P_a inside cavitation bubble as function of gas-content of bubble, expressed by ratio P/P_a.
P_a: static pressure exerted on the liquid P: gas pressure inside the bubble at the beginning of collapse [Ref. (*29*)]

the shock wave pressure set up by the collapse of a cavitation bubble is governed by several factors, such as the damping action of gases and vapor inside the bubble, the viscosity and compressibility of the liquid, the finite speed of collapse, the shape of the bubble and finally the break-up of the bubble during collapse, which leads to the formation of a variety of smaller bubbles. Many experimental investigations on shock waves radiated from collapsing cavitation bubbles have been carried out using cavities generated by electrical spark discharges or by the mechanical deceleration of fast-moving liquid columns. High-speed cinematographic investigations of Schmid (*82*) on the collapse of isolated cavities having an initial radius up to 1 cm, demonstrated that a very strong shock wave of approximately spherical shape is radiated from the cavity at the moment of implosion. The shock waves generated in this way were so strong that they were visible even without the use of schlieren optics. Investigations of Mellen (*51*) on shock waves radiated from the cavities generated by spark discharges or exploding wires, indicate that the maximum radial velocity of the cavity wall is of the order of the velocity of sound in the liquid.

Bohn (*7*) recorded the pressure and the sound spectra of cavitation generated by a 15 kc/sec ultrasonic transducer. In the pressure recording obtained, sharp peaks caused by the collapse of cavitation bubbles are superimposed on the normal sound spectrum. It is, however, impossible from these results to determine the peak pressure of cavitation collapse directly, since the shock wave impinges only on a very small fraction of the total area of the hydrophone employed and the signal obtained is integrated over an unknown area. The estimated values for the peak pressure inside a cavity are of the order of 10^3 bars.

Mundry and Güth (*30*) photographed the shock waves generated by ultrasonic cavitation at 2.5 kc/sec by using schlieren optics. Their experiments demonstrate that shock waves are not generated periodically but appear at irregular intervals. Moreover, the centers of the shock waves were always in the proximity of the ultrasonic transducer interface, indicating that only in this region are cavitation bubbles large enough to generate shock waves at their implosion. It has also been confirmed experimentally that the radiated shock waves are more intense the lower the gas content of the cavitation bubble.

Since the amplitude of the shock waves generated by cavitation cannot be measured directly, because the hydrophones available today are much too large compared to the dimensions of the cavity, the experimental determination of the pressure must be carried out at comparatively great distances from the cavity. The peak pressure due to the collapse of spark-induced cavities recorded by Mellen (*51*) on an electroacoustic hydrophone at 100 cm from the cavity was in the range of 0.5 to 5 bar. The amplitude of the shock wave, however, is considerably higher because it is strongly attenuated even over moderate distances. Most investigations indicate that the maximum pressure drops inversely proportional to the distance travelled by the shock wave (*20*). Measurements of the peak pressure at relatively short distances from the center of the cavity, as well as the determination of the plastic deformation of metals with known tensile strengths, reveal that the pressures generated by collapsing cavities fall within the range of 3×10^3 and 1.3×10^4 bars (*20*). From these results it becomes clear that the shock waves radiated from collapsing cavities can produce very drastic effects within the irradiated liquid.

3. Physicochemical Actions of Ultrasound on Macromolecules

3.1. Chemical Effects

In most cases the propagation of ultrasonic waves in liquids can be regarded as an adiabatic process. The changes in pressure and consequently in density in the vibrating liquid are so rapid that heat cannot be transferred from the compressed region to the surrounding medium and vice versa from the surrounding medium to the rarefied regions of the liquid. At the passage of a sound wave through a liquid the temperature difference between the vibrating layers due to adiabatic compression is of the order of hundreths of a degree only. No chemical reactions are to be expected from this. In the cases where ultrasonic irradiation accelerates certain reactions, this is mainly caused by microstirring of the reagents, which greatly enhances the mass transfer, or by the physical destruction of boundary layers or protective films in the case of heterogeneous reactions.

Chemical effects in an ultrasonic field are produced exclusively by cavitation or vibrating resonance bubbles. The compression of the gas present in the bubble generates high temperatures inside, which are responsible for many chemical processes occurring in an ultrasonic field. Moreover, at the initial stages of formation of a cavitation bubble, electrical charges may appear. These charges are not uniformly distributed over the walls of the bubble and electronic breakdown may occur. This is accomplished by the emission of visible and ultraviolet light and ionization of molecules present in the cavity can occur. All these effects depend greatly on the nature of the gas present in the cavitation bubble. Chemical effects play an important role in aqueous media, because being a favorable medium for the occurrence of cavitation, water additionally ensures electronic breakdown of the cavitation bubbles. Moreover, reactions due to ultrasonics occur in organic solvents too, especially if traces of water are present.

The following discussions are limited to those reactions that are associated in some manner with ultrasonic depolymerization. The decomposition of water is accompanied by the formation of radicals and hydrogen peroxide, especially in the presence of oxygen. The decomposition of water is considered to be initiated by the ejection of an electron from the water molecule due to the action of ionizing radiation, followed by a series of secondary processes, such as:

$$
\begin{array}{llll}
H_2O & \longrightarrow & H_2O^+ & + e^- \\
H_2O^+ & \longrightarrow & H^+ & + OH^\cdot \\
H^+ + e^- & \longrightarrow & H^\cdot & \\
OH^\cdot + OH^\cdot & \longrightarrow & H_2O_2 & \\
H^\cdot + H^\cdot & \longrightarrow & H_2 & \\
H_2O & \longrightarrow & H^\cdot & + OH^\cdot \\
H^\cdot + O_2 & \longrightarrow & HO_2^\cdot & \\
HO_2^\cdot + HO_2^\cdot & \longrightarrow & H_2O_2 & + O_2 \qquad \text{etc.}
\end{array}
$$

The existence of all these free radicals has been demonstrated experimentally. Further details on this subject are given by Él'Piner (*17*) and Nosov (*67*).

Decomposition of organic halogen derivatives occurs readily if they are irradiated in aqueous media. If carbon tetrachloride is used, atomic chlorine is split off the molecule, even in the total absence of oxygen. The chlorine liberated can easily be detected with potassium iodide and a starch indicator. The production of iodine from solutions of potassium iodide containing carbon tetrachloride is therefore often used as a standard test for the intensity of cavitation.

The different radicals formed during ultrasonic irradiation of aqueous solutions can affect degradation studies, since the products formed can react with polymers in several ways. Oxidation reactions due to the $OH^{\cdot}$ and the $HO_2^{\cdot}$ radicals play an important role in the degradation of polymethacrilic acid at 250 kc/sec. Hydrogen peroxide, however, which is also formed, has no effect on the polymer (*1*). The chemical effects of ultrasound on degradation experiments can be avoided by the addition of certain protective substances, which act as radical scavengers, such as allyl thiourea, iodine or even stable free radicals, such as α, α'-diphenyl-β-picryl-hydrazyl (DPPH). Since cavitation bubbles behave almost isothermically at low frequencies and adiabatically in the megacycle range, high temperatures and ionization will not occur in the cavities at low frequencies. Consequently chemical effects may be disregarded in most cases at frequencies near 20 kc/sec. In view of this, low frequency ultrasonics should be used as far as possible for the investigation of degradation reactions of polymers in aqueous media.

3.2. Depolymerization

Most macromolecular compounds in solution suffer deep alterations when they are exposed to high-intensity ultrasonics. The most important actions of ultrasound consist of the liquefaction of thixotropic gels and the depolymerization of macromolecules. Many investigators found a reversible reduction of the viscosity after insonation of several naturally occurring high polymers, such as gelatin, starch, agar-agar and gum arabic, which could be explained assuming that ultrasound caused the dispersion of aggregates, rather than the breakage of chemical bonds. For most polymers, however, a definite effect of depolymerization is observed, especially at high ultrasonic intensities. Examples are provided by solutions of polystyrene, polyvinyl acetate, polyacrylates, nitrocellulose, enzymes, DNA, dextran and many other polymers. Reviews of the earlier investigations have been given by Bergmann (*5*) and Wilke and Altenburg (*107, 108*).

Schmid and Rommel (*83*) were the first to observe an irreversible reduction in the viscosity of solutions of polystyrene, polyacrylates and nitrocellulose due to the breakage of covalent bonds in the polymer chain. These investigators also found that in the first minutes of irradiation depolymerization was fairly rapid, but slowed down subsequently and ceased altogether when a minimum molecular weight was approached. The existence of a limiting degree of polymerization, beyond which no degradation occurs, was also established by other investigators (*17, 92*) and constitutes the basis of most degradation mechanisms. A correlation of the limiting degree of polymerization with the microstructure of the polymer has also been suggested (*93–95*).

The depolymerizing effect of ultrasound is more pronounced for higher molecular weights of the polymer. This relationship between degradation rate and chain length has been found by all investigators. The rate of degradation depends mainly on the duration of the treatment, on the concentration of the solution, on the nature of the polymer and solvent, and the intensity of ultrasound. From all experiments there is evidence of nonrandom scission of the polymer molecules, with clear preference of rupture near the midpoint of the chain. No fundamental differences have been reported in the mechanism of degradation of synthetic polymers in organic solvents (*24, 76, 97, 103*) and naturally occurring polymers or proteins in aqueous media (*2, 3, 9, 16, 78, 81*). A detailed discussion of all the parameters that affect ultrasonic degradation will be given in Chapter 4.

Many attempts have been made to develop mechanisms to explain degradation. Since chemical effects play a minor role, the mechanical nature of ultrasonic degradation is nowadays generally accepted. Depolymerization is caused by the mechanical forces that arise in the liquid due to the propagation of acoustic energy through it. Moreover, the major part of these forces are directly associated with cavitation and the formation of strongly vibrating bubbles in the liquid treated. The results of most investigations led to the conclusion that cavitation is responsible for depolymerization (*17, 105, 106*). Strong velocity gradients appear in the liquid close to collapsing cavitation bubbles and may cause breakage of the macromolecules; besides this, shock waves also play an important role. By applying an external pressure on the liquid in order to prevent cavitation, it was found that the rate of degradation was greatly reduced (*17*). In carefully degassed liquids or in liquids saturated with carbon dioxide, where cavitation is almost completely suppressed, no depolymerization was detected.

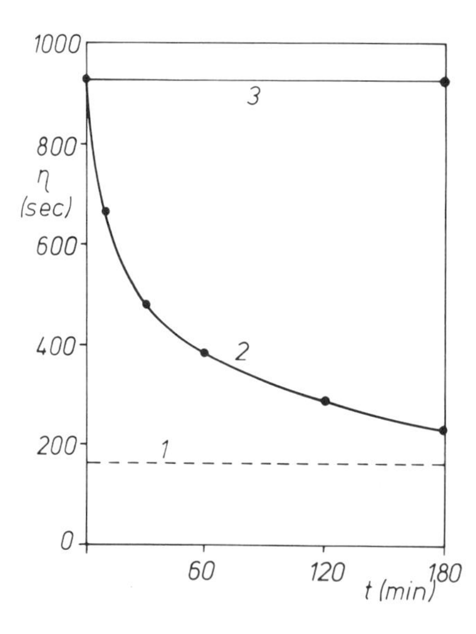

Fig. 11. Effect of cavitation on ultrasonic degradation of polystyrene in toluene at 400 kc/sec and power of 50 watts. Viscosity of solution (η – given by flow-time in viscometer) is plotted against time of irradiation (t). *1*: pure toluene *2*: solution, air present *3*: solution degassed [Ref. (*105*)]

The role of cavitation was clearly demonstrated by Weissler (*105, 106*), who investigated the degradation of polystyrene in toluene and that of hydroxyethyl cellulose in water at an ultrasonic frequency of 400 kc/sec. The results for polystyrene are shown in Fig. 11. During the irradiation of the first sample many cavitation bubbles were present and decrease in molecular weight, as measured by viscosity, occurred to about one tenth of the initial value. The second sample, carefully degas-

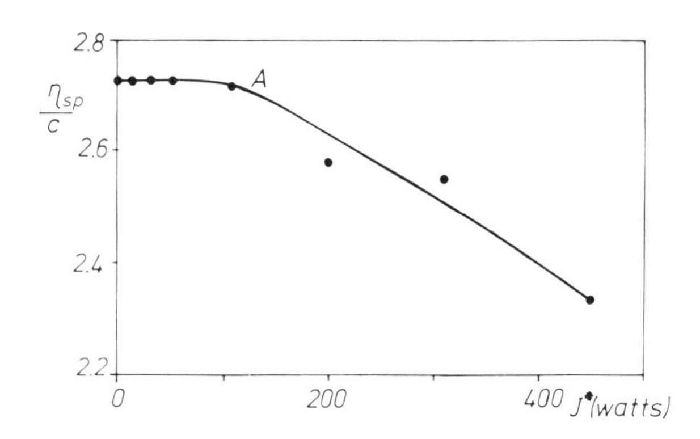

Fig. 12. Effect of ultrasonic power (J^*) and cavitation on degradation of hydroxyethyl cellulose. Ordinates are reduced viscosity of solution. Irradiation time: 5 min at 400 kc/sec. At point A cavitation set on [Ref. (*105*)]

sed by boiling under vacuum, showed no cavitation bubbles during irradiation and suffered no appreciable change in molecular weight. An even more direct relation between cavitation and degradation is demonstrated in Fig. 12. In this case samples of hydroxyethyl cellulose were subjected to ultrasonic irradiation at increasing intensity levels. No depolymerization occurred at low ultrasonic intensities, when cavitation was absent; at the ultrasonic intensity where cavitation began, depolymerization took place and became greater as the intensity was further increased.

Some investigators, however, claim to have detected degradation in the absence of cavitation, provided the molecular weight of the starting material was high enough (*52*). At very high frequencies, in the megacycle range and in the absence of cavitation, a very slight degradation seems possible, according to Mostafa (*60*), most probably due to some kinds of resonance effects. It must be noted, however, that at high frequencies and with the high ultrasonic intensities then required to have measurable effects, it is very difficult to establish the experimental conditions where cavitation is thoroughly suppressed. The degradation found by earlier investigators in the "absence" of cavitation must be therefore attributed in most cases to their inadequate method for eliminating cavitation.

3.3. Formation of Macroradicals

The mechanical rupture of polymer chains leads to the formation of free valences on the ends of the chain fragments. According to Henglein (*33*) there are three basic possibilities for the depolymerization reaction. In the so-called *homolytic cleavage* two free macroradicals are formed:

$$P_{n+m} \longrightarrow P_n^{\cdot} + P_m^{\cdot}$$

As an example there is the rupture of polystyrene:

$$\text{---}CH_2\text{--}\underset{C_6H_5}{CH}\text{--}CH_2\text{--}\underset{C_6H_5}{CH}\text{---} \longrightarrow \text{---}CH_2\text{--}\underset{C_6H_5}{CH^{\bullet}} + {}^{\bullet}CH_2\text{--}\underset{C_6H_5}{CH}\text{---}$$

In the *heterolytic cleavage* two macromolecular ions with opposite charge appear:

$$P_{n+m} \longrightarrow P_n^+ + P_m^-$$

As an example for this reaction the heterolytic cleavage of polydimethylsiloxane in the presence of methanol may be cited (*104*). Methanol, a strong nucleophile, reacts very rapidly with carbonium and siliconium ions, thus indicating the heterolytic cleavage of the Si-O bond:

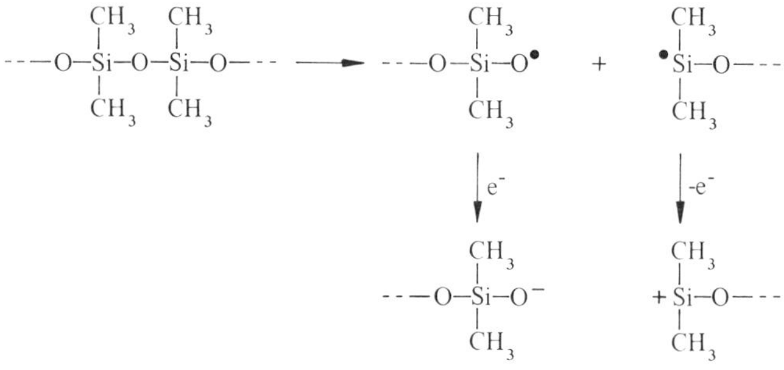

$$\begin{array}{c} CH_3 \quad CH_3 \\ | \qquad | \\ ---O-Si-O-Si-O--- \\ | \qquad | \\ CH_3 \quad CH_3 \end{array} \longrightarrow \begin{array}{c} CH_3 \\ | \\ ---O-Si-O^{\bullet} \\ | \\ CH_3 \end{array} + \begin{array}{c} CH_3 \\ | \\ {}^{\bullet}Si-O--- \\ | \\ CH_3 \end{array}$$

$$\begin{array}{c} \downarrow e^- \\ CH_3 \\ | \\ ---O-Si-O^- \\ | \\ CH_3 \end{array} \qquad \begin{array}{c} \downarrow -e^- \\ CH_3 \\ | \\ +Si-O--- \\ | \\ CH_3 \end{array}$$

The third possibility is the intramolecular *disproportionation* with the formation of two stable macromolecular fragments:

$$P_{n+m} \longrightarrow P_n + P_m$$

The degradation of polydimethylsiloxane in the absence of methanol is an example considered to follow this mechanism:

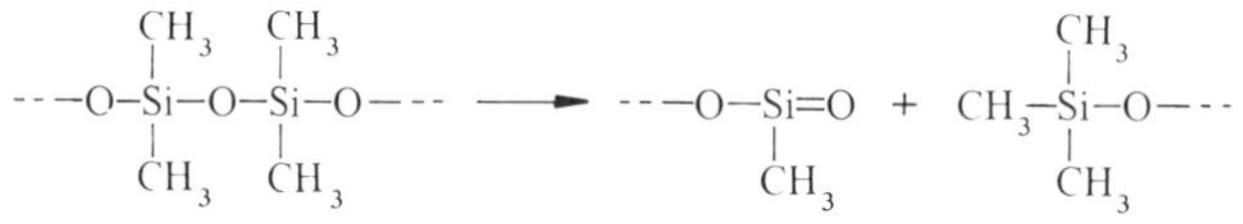

$$\begin{array}{c} CH_3 \quad CH_3 \\ | \qquad | \\ ---O-Si-O-Si-O--- \\ | \qquad | \\ CH_3 \quad CH_3 \end{array} \longrightarrow \begin{array}{c} \\ \\ ---O-Si{=}O \\ | \\ CH_3 \end{array} + \begin{array}{c} CH_3 \\ | \\ CH_3-Si-O--- \\ | \\ CH_3 \end{array}$$

Experimental investigations have shown that the breakage of the C–C bond by the action of ultrasound usually leads to the formation of long-chain radicals. Methanol does not react with the degradation products formed, which indicate the homolytic nature of the rupture of the C–C bond. The macroradicals formed can be inactivated by recombination or by disproportionation; since the number of macroradicals formed during degradation of dilute solutions is small compared to the number of solvent molecules, recombination and disproportionation are usually repressed by the reaction with solvent molecules or other substances present in the solution. The degradation products formed during the ultrasonic irradiation of polymers react preferentially with substances capable of inactivating free radicals, such as iodine, oxygen or allyl thiourea. If oxygen is present in the solution, only a small fraction of the macroradicals formed react with iodine, which means that oxygen is more reactive.

The quantitative detection of free radicals formed during ultrasonic degradation is best carried out by adding stable free radicals to the solution, α,α'-diphenyl-β-picrylhydrazyl (DPPH) being most commonly used. Henglein (*33, 34*) showed that the reduction in DPPH concentration, which can be measured photometrically, was proportional to the number of C–C bonds broken during the irradiation of solutions of polystyrene and polymethylmethacrylate. He also demonstrated that two molecules of DPPH are used per ruptured bond. Not all DPPH molecules that react with the macroradicals, however, are incorporated into the polymer chains, a disproportionation reaction with transfer of an hydrogen atom from the polymer to the DPPH molecule taking place.

The free radicals formed by the irradiation of polymer solutions can also start polymerization reactions, if monomers are present. It has been shown that the rate of polymerization increases with increasing polymer concentration, since more macroradicals are then formed. As soon as the degree of polymerization exceeds the limiting chain length for degradation, these polymers may also undergo mechanical degradation. Therefore, this reaction may be regarded as autocatalytic. Examples of the degradation and mechanochemical polymerization of several polymer–monomer systems under ultrasonic irradiation are given by Henglein (*35*), Él'Piner (*17*) and Fujiwara *et al.* (*21*).

4. Parameters Affecting the Degradation of Polymers in Solution

4.1. Quantities Characterizing Ultrasound

In order to explain the depolymerizing action of ultrasonic waves, it is necessary to take into consideration all the factors that are known to have an influence on the degradation process. Since the numerous effects that result from the collapse of cavitation bubbles are responsible for the scission of long-chain molecules, the influence of the quantities characterizing the ultrasonic field on the degradation reaction is in many aspects the same as their influence on cavitation. Among the series of factors that affect cavitation, the most important are the acoustic intensity, the frequency of ultrasound, and the static pressure exerted on the liquid. The acoustic intensity defines the size of the zone of liquid where cavitation occurs and controls the probability of occurrence of cavitation events per unit volume. The frequency of ultrasound determines the half period of collapse of the cavitation bubbles and establishes the resonant bubble radius and the probability of cavitation events to occur per unit time. Finally, the static pressure defines the size distribution of cavitation nuclei, from which a series of variables are dependent, such as the number of cavitation events, cavitation threshold, velocity of bubble growth and collapse and the efficiency of energy transfer.

The *ultrasonic intensity* defines the acoustic pressure amplitude, which determines the threshold necessary to produce cavitation. As the acoustic pressure amplitude is increased, both the number of bubbles and their maximum size increases, resulting in an increased overall cavitation activity. Okuyama (*69*) demonstrated that the intensity of collapse of a cavitation bubble does not strongly depend on the ultrasonic intensity. The main effect of increasing ultrasonic intensity is that a larger number of cavitation bubbles are formed. Therefore, to alter cavitation intensity over wide ranges, the hydrostatic pressure over the liquid or the temperature must be changed. Mostafa (*57*) and Jellinek (*45*) showed that the degradation rate constants of polystyrene in benzene increased almost linearly over a fairly large range of ultrasonic intensities, provided the ultrasonic intensity was higher than the threshold value for cavitation. Their results are shown in Figs. 13 and 14.

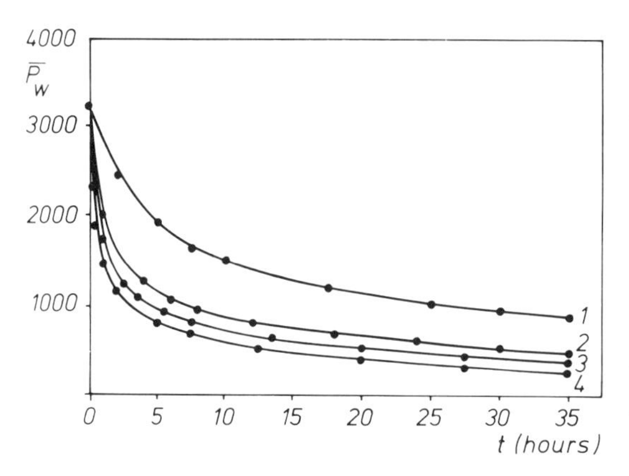

Fig. 13. Effect of ultrasonic intensity on degradation of polystyrene: variation of weight-average chain length ($\bar{P}_w$) with time of irradiation (t). *Curve 1:* 4.9 watts/ cm^2 *2:* 9.6 watts/cm^2 *3:* 12.5 watts/cm^2 *4*: 15.8 watts/cm^2 [Ref. (*57*)]

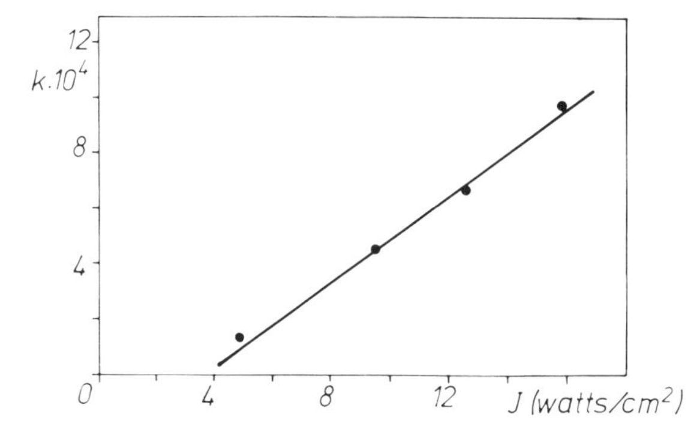

Fig. 14. Initial rate constants of degradation (k) as a function of the intensity of ultrasound (J). [Ref. (*45*), calculated using data from Fig. 13]

According to the results of most investigators, the limiting degree of polymerization seems to decrease with increasing ultrasonic intensity. The experiments of Okuyama (*69*) and Thomas and Alexander (*102*), however, indicate that the degradation limits of polymerization should be nearly independent of ultrasonic intensity, the time of irradiation being much more important. This is a factor of uncertainty occurring in most of the earlier investigations, in which the degradation curves always showed a tendency toward lower degrees of polymerization, even after the longest periods of irradiation used by these researchers. Consequently the "limiting" degrees of polymerization which they found were nearly always several times larger than the values obtained more recently using gel permeation chromatography. In addition to this, viscosity-average molecular weights, which were invariably used, are not suited to the determination of very low molecular weights.

In order to relate degradation time to ultrasonic intensity some investigators suggested that degradation should be correlated on the basis of the product of ultrasonic intensity with degradation time, *i. e.*, the total energy input per unit of mass of the polymer irradiated (*18, 96*). This relation, however, is only valid over a comparatively small range because the total input of energy in the system greatly exceeds that fraction of energy necessary to break the chemical bonds. Schmid (*84*) reported that in his system, only 4 bonds were broken from 10^7 molecules present in the solution in every period of ultrasonic oscillation. The energy fraction from the total input energy, which causes rupture of chemical bonds is of the order of 10^{-5} or less.

There is no possibility of isolated polymer molecules being directly excited to resonance vibrations due to ultrasound. The applied ultrasonic frequencies, at most 2000 kc/sec, are by many orders of magnitude smaller than the fundamental frequencies of vibration of chemical bonds (10^{13}–10^{14} c/sec). The wavelength of ultrasound in liquids is of the order of 1 mm for the highest frequencies used in practice, therefore, macromolecules that usually have contour lengths of 0.1–1×10^{-3} mm, can never get into different regions of an ultrasonic wave, where they might be submitted to dragging forces. Consequently the *ultrasonic frequency* has no direct effects on degradation, but again, cavitation thresholds are strongly dependent on frequency. As the acoustic frequency is increased, the radius of cavitation bubbles decreases, resulting in an increase in the cavitation threshold. An increase in frequency also means shorter acoustic periods, lower maximum bubble size and thus, less cavitation intensity. Jellinek (*45*) deduced an expression in terms of characteristic quantities of the polymer and the cavitation bubbles, finding that the degradation rate should be constant for frequencies below 500 kc/sec. This is in agreement with Schmid and Poppe's (*89*) experimental results, that at an acoustic intensity of 1 watt/cm^2, degradation constants were independent on frequency in the range from 10 kc/sec to 300 kc/sec.

At very high frequencies, in the megacycle range, Mostafa (*59*), however, found a marked dependence of the rate of degradation on the ultrasonic frequency. The change in the weight-average chain length of polystyrene in benzene with time of irradiation is shown for different frequencies and an intensity of 12.5 watts/cm^2 in Fig. 15. Degradation increased from 750 kc/sec until it reached a maximum value at 1000 kc/sec, where cavitation was also maximum, decreased thereafter continuously as the frequency approached 2000 kc/sec, a point where cavitation was almost suppressed, showing in this way that cavitation played the major role in the process of degradation.

In order to decide whether the effects observed were due to changes in cavitation intensity or due to a direct action of the different ultrasonic frequencies on the polymer molecules, Mostafa degraded the same sample at 1000 kc/sec, where degradation was maximum at atmospheric pressure, but prevented the formation of cavities by irradiating the solution under vacum. He obtained a slight degradation, similar to that obtained at 2000 kc/sec in the presence of air. From this result he concluded that high polymers can be ruptured by ultrasonic waves directly in the megacycle range, provided the acoustic intensity was high enough and the experimental conditions permitted resonance effects to take place in the entangled polymer molecules.

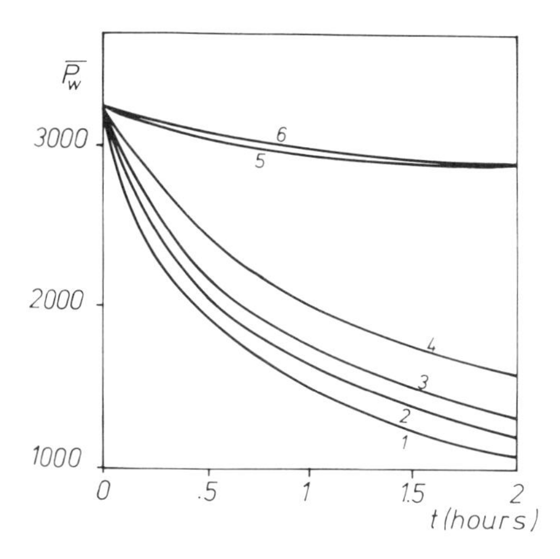

Fig. 15. Effect of ultrasonic frequency and cavitation on degradation of polystyrene: variation of weight-average chain length ($\bar{P}_w$) with time of irradiation (t). *Curve 1:* 1000 kc/sec *2:* 1250 kc/sec *3:* 750 kc/sec *4:* 1500 kc/sec *5:* 1000 kc/sec *6:* 2000 kc/sec. Curves *1, 2, 3, 4* and *6* at atmospheric pressure; curve *5* under vacuum. Ultrasonic intensity = 12.5 watts/cm^2 [Ref. (*87*)]

The maximum radius of a cavitation bubble is also a function of the hydrostatic pressure exerted on the system; within the threshold limits it decreases linearly with increasing hydrostatic pressure. On increasing the *static pressure* over a liquid, some of the gas present in the cavitation bubble dissolves in the surrounding liquid, thus further diminishing its radius. Therefore, these effects tend to decrease cavitation intensity. Increasing hydrostatic pressure can, however, also increase the amount of energy that is transferred from the ultrasonic transducer to the liquid, due to better contact at the solid-liquid interface, increasing degradation effects. Investigations of Jellinek and Brett (*43, 45*) showed that degradation at first increased with pressure and subsequently decreased as pressure was further increased. This is understandable, since cavitation is more and more suppressed as static pressure is increased. Appreciable differences were found in the degradation rate if pressure was exerted on the solution by a gas or directly by means of a mercury column. In either case the rate constants for degradation passed a maximum as the pressure was increased. For gaseous pressure a broad maximum was observed and the decrease of the rate constants with further increase of pressure was gradual, showing a tendency to reach a constant value. In these experiments the degradation rate never fell to zero, even after the pressure exerted on the liquid was 6 times larger than the acoustic pressure amplitude. This fact is undoubtedly in part due to the greater solubility of the gas in the liquid at higher pressures. For pressure exerted by means of a mercury column the maximum was more pronounced and the rate constants decreased rapidly to zero as the pressure was further increased. The pressure necessary to suppress cavitation completely was in this case about 6 times the value of the acoustic pressure amplitude. These results show clearly that increasing of static pressure over the solution is a very inefficient way of suppressing cavitation.

4.2. Properties of the Solution

The propagation of ultrasonic waves in liquids and the phenomena resulting from this depend greatly on the properties of the liquid. Since ultrasound is transmitted differently by the pure liquid and the dissolved polymer molecules, the first question that arises, is to find out if the *density* of the solution, or more precisely, the difference between the densities of the solution and that of the polymer coil, affects degradation. If inertial forces resulting from the relative movement between the molecules of the solvent and the macromolecules are supposed to be responsible for degradation, the density of the solvent should have a clear effect on the degradation rate and the limiting degree of polymerization. Schmid and Beuttenmüller (*87*) carried out such investigations. In order to alter the density of the solvent, these investigators prepared different toluene-carbon tetrachloride and cyclohexane-carbon tetrachloride mixtures and degraded polystyrene in these mixed solvents. Furthermore, they prepared a solvent mixture having exactly the same density as the polymer to be investigated. The results are shown in Fig. 16. These experiments demon-

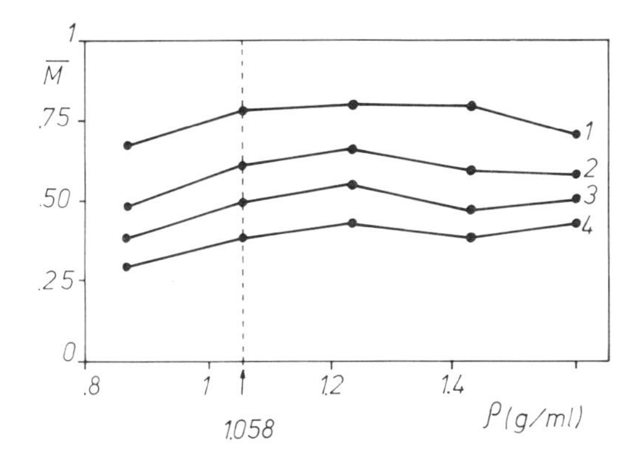

Fig. 16. Effect of density of solvent (ρ) on degradation of polystyrene. Ordinates are mean molecular weights ($\overline{M}$) given as fraction of initial molecular weight. Parameter of curves is time of irradiation: *1:* 10 min *2:* 30 min *3:* 60 min *4:* 120 min. Solvent is mixture of toluene and carbon tetrachloride [Ref. (*87*)]

strate clearly that within experimental errors, degradation is not affected by the density of the solvent. Consequences resulting from these facts on the mechanism of degradation will be discussed in another chapter. The direct thermal effects of ultrasound are known to be not responsible for the degradation of polymers. An influence of *temperature* on degradation must therefore be related to changes in cavitation intensity. Schmid and Beuttenmüller (*88*) investigated the degradation of nitrocellulose dissolved in n-butylacetate and of polystyrene in toluene at temperatures ranging from 40 °C to 120 °C. Their results, corrected for thermal degradation, show that degradation is less pronounced at higher temperatures. The limiting degree of polymerization was found to be considerably higher at high temperatures. Similar effects were found by Thomas and Alexander (*102*), who degraded cellulose nitrate in a series of alkyl acetates as solvents, in the temperature range from 0 °C to 85 °C. In each solvent a temperature range was found, where degradation was maximum. In this temperature range cavitation intensity was also maximum. This shows that temperature affects degradation by the magnitude of the vapor pressure inside the cavitation bubbles. This fact is easily understandable, because if vapor pressure of the solvent becomes appreciably high, solvent molecules evaporate into the cavitation bubble during their growth and cushion the shock effect in the subsequent stages of collapse.

During the collapse of cavitation bubbles *different gases* inside the bubbles develop different temperatures and pressures for equal compression ratios because of the different specific heat ratio C_p/C_v of each gas. The effect is more important the higher the frequency of ultrasound, because under these conditions the collapse of a cavity can be regarded as an adiabatic process. If the gas present in the bubble is monoatomic a much higher final temperature will be reached, as in the cases of diatomic or polyatomic gases. Therefore, if ultrasonic degradation were thermal in origin, appreciable differences in the rate of degradation should be found. The investigations of Jellinek and Brett (*42*), however, showed that even at an ultrasonic frequency of 500 kc/sec no significant differences could be detected, if polystyrene was degraded in benzene saturated with air or argon. Very slight effects were indeed found by Melville and Murray (*52*), who saturated their solutions with air, oxygen and nitrogen.

The degradation rate depends, however, markedly on the *solubility* of the gas in the liquid. If the solubility of the gas is great, a larger quantity will enter the cavitation bubbles during their expansion, and exert a cushioning effect during the collapse, in this way considerably diminishing the intensity of the shock wave radiated at the final stages of collapse. This is confirmed by Jellinek's experimental results, in which it was found that the rate of degradation is appreciably reduced with increasing solubility of the gas.

A similar effect is exerted by the solvent itself. If the solvent has a high volatility, the quantity of vapor, which enters the cavitation bubble during expansion, is large. A strong cushioning effect in the final phase of the collapse is then to be expected, so that the intensity of the shock wave is reduced. The *volatility* of liquids can be expressed by its enthalpy of vaporization. The higher the enthalpy of vaporization of the solvent, the higher is the rate of degradation since in this case less vapor is present in the cavitation bubble, and only a very slight cushioning effect is possible. Therefore, a very intense shock wave is radiated from the collapsing cavity, resulting in a high degradation rate. The investigations of Basedow and Ebert (*2*) showed that these considerations are in accordance with the experimental facts. For dextran in a series of solvents an almost linear relationship between the degradation rate constant and the enthalpy of vaporization of the solvent was found (Fig. 17).

The opposite effect is found if diethyl ether is present in the solution. Because of its high volatility the collapse of cavitation bubbles is considerably damped, resulting in a decrease of the rate constant. Basedow and Ebert found that the rate of degradation of dextran in water was decreased to one tenth of its initial value, if the solution was saturated with diethyl ether.

The effects of *viscosity* and *surface tension* of the solution on the degradation process are not pronounced. Jellinek (*44*) extended the theory of the collapse of cavitation bubbles, including the effects of viscosity and surface tension. From his observations it follows that high surface tension accelerates the collapse of the cavities, whereas high viscosity has a retarding effect. According to Okuyama (*69*), however, increasing viscosity retards only the diffusion of gas molecules from the

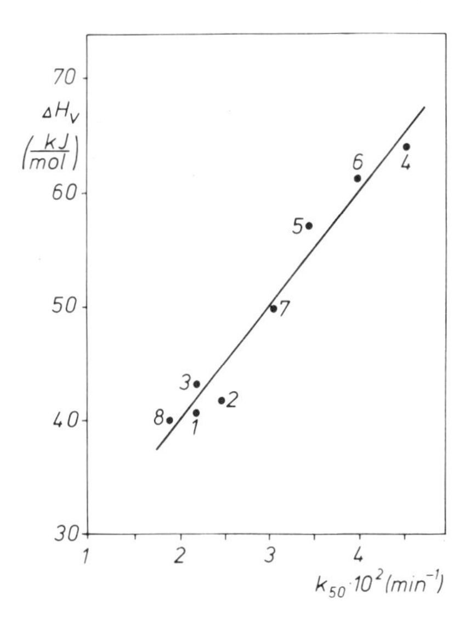

Fig. 17. Dependence of degradation rate constant of dextran with molecular weight of 50000 (k_{50}) on enthalpy of vaporization of solvent (ΔH_V). Numbers represent solvents: *1:* water *2:* deuterium oxide *3:* dimethylsulfoxide *4:* formamide *5:* ethylene glycol *6:* glycerol *7:* ethanolamine *8:* methanol 20% in water. Ultrasonic intensity 24 watts/cm^2, frequency 20 kc/sec [Ref. (*2*)]

liquid into the cavities, thus diminishing the cushioning effect in the final stage of collapse and increasing therefore the rate of degradation. The overall effect of surface tension and viscosity may therefore be disregarded in most cases. This is in accordance with the experimental results of Basedow and Ebert (*2*), who did not find alterations in the degradation rate constants for dextran in water and water with a surfactant, due to surface tension, or dextran in ethylene glycol and glycerol, due to viscosity effects.

Since ultrasonic degradation of polymers is due to mechanical effects, the *configuration* of the molecule would be expected to have an influence on the degradation rate. Alexander and Fox (*1*) investigated the degradation of polymethacrylic acid in solutions of different ionic strengths and showed that the degradation rate increased as the molecule became more asymmetric. Schmid and Beuttenmüller (*87*) found in agreement with this that the addition of a nonsolvent to the solution, in their case acetone to polystyrene in toluene, decreased the rate of degradation, thus confirming that poorly solvated polymer molecules are more difficult to degrade. Suspensions of a polymer in a nonsolvent are not degraded at all. Similar results were obtained by Basedow and Ebert (*2*), who investigated the degradation of dextran in a series of different solvents and aqueous media, showing that the rate of degradation increased as the polymer coil became more expanded and decreased in poor solvents or if a nonsolvent was added to the solution. The alterations in the rate constants were, however, small. There is also evidence that the effect of polymer solvation on the degradation rate is dependent on the nature of the polymer investigated. Thus, Thomas and Alexander (*102*) did not find alterations in the rate constants for degradation of cellulose nitrate in ethyl acetate, in the presence of variable quantities of ethanol as a nonsolvent or cellulose nitrate in acetone, using variable quantities of water as a nonsolvent.

5. Experimental Investigation of Degradation

5.1. Kinetic Analysis of the Degradation Reaction

Degradation of polymers by the action of ultrasound is normally characterized by a series of parallel-proceeding reactions, since the original substances usually have a more or less broad molecular-weight distribution. The exact rate constants of degradation are therefore not obtainable directly from kinetic studies because of their dependence on molecular weight. Exact rate constants are only obtainable if the original substance is monodisperse or if certain assumptions concerning the degradation mechanism are made. The fact that the molecular-weight distribution of a polymer changes throughout degradation, makes a complete kinetic analysis in terms of molecular constant very difficult.

Assuming random scission and that the rate of degradation decreases progressively with decreasing chain length, and reaches zero at the limiting degree of polymerization, Jellinek and White (*38*) calculated the number and weight distribution functions and the number- and weight-average chain lengths of an initially monodisperse polymer sample, after a certain time of ultrasonic irradiation. Using matrix algebra, Mostafa (*54, 55*) presented a more general and exact solution of the rate equations representing the degradation of polymers. The assumptions made are the same as Jellinek's, but Mostafa considered the degradation of monodisperse polymers as well as polymer samples with a definite molecular-weight distribution. Ovenall and Hastings (*72, 74*) derived relationships for the rate of degradation of polymers assuming that molecules only degrade, if the resulting fragments are greater than $P_e/2$ where P_e is the limiting degree of polymerization, and that all bonds in the chain are equally likely to break, except those within $P_e/2$ monomer units from each end. If dB_i/dt is the rate of breakage of molecules having a degree of polymerization P_i, the proposed rate equations are:

$$\frac{dB_i}{dt} = k\,(P_i - P_e)\,n_i \quad \text{for } P_i > P_e \tag{11}$$

$$\frac{dB_i}{dt} = 0 \quad \text{for } P_i \leqslant P_e \tag{12}$$

where k is the rate constant, which is assumed to be independent of P_i, and n_i is the number of molecules having a degree of polymerization P_i. This means that if a molecule is large enough to be ruptured, the probability of a bond being broken in the degradable section is proportional to the number of bonds in this section, whereas the section of the bond broken is random.

Many of the assumptions which formed the basis of these earlier kinetic calculations are not, however, in accordance with the experimental facts obtained more recently using gel permeation chromatography. Therefore, all these calculations are of very limited practical importance.

Based on experimental results, Schmid (*84*) concluded that the rate of degradation dB/dt of a molecule with a degree of polymerization of P, was proportional to that fraction of the total chain, which exceeded the limiting degree of polymerization P_e, *i.e.:*

$$\frac{dB}{dt} = k\,(P - P_e) \qquad (13)$$

where k is a general rate parameter which depends on the polymer system under investigation. Moreover, he found that the rate constant of degradation depends on the concentration of the polymer and the chain length of the molecule.

Investigations of Schmid and Rommel (*83*), Thomas and Alexander (*101*), Gooberman and Lamb (*27*) and Jellinek and White (*41*) demonstrated that the degradation rate rapidly decreased as the *concentration* was increased. In some cases (*41, 27*) a maximum in the rate constants has been found, *e.g.,* for polystyrene in benzene, at the concentration where the polymer chains began to be entangled. The effect of concentration is certainly a very complex one, since entanglements and non-Newtonian fluid flow due to cavitation cannot be described by simple mathematical relations. Earlier investigators claimed that degradation would cease in very dilute and in highly concentrated solutions. These results were, however, erroneous, since it has been found later that degradation proceeds quite normally at infinite dilution, in very concentrated solutions and even in cross-linked gels.

All experimental investigations on ultrasonic degradation of polymers demonstrated that the rate constants increased with increasing chain length. Furthermore, Mostafa (*58*) showed that the limiting chain length is independent of the initial chain length, provided the intensity of ultrasound is kept constant and degradation time is long enough. Jellinek (*45*) showed that for polystyrene in benzene the dependence of the degradation rate constant was almost linear on the degree of polymerization. The complete kinetic analysis requires, however, the knowledge of the molecular-weight distribution after different periods of irradiation.

Starting from the molecular-weight distribution of the original polymer, both the molecular weight and the concentration are altered during degradation. This is represented in a three-dimensional molecular weight–concentration–time diagram in Fig. 18. The variation in concentration of polymer molecules with definite molecular weights is represented by planes parallel to the *c-t*-plane, which intercepts the molecular weight axis. It is clearly shown that the degradation curves concentration–time are different for the different molecular weights. In the range of high molecular weights, only degradation takes place and the concentration of these molecules decreases monotonously with time. At small molecular weights, no molecules are initially present; the degradation of larger molecules, therefore, causes their concentration to rise at first. After a certain period of time, degradation of the small molecules is faster than the formation, due to the degradation of larger ones, and the concentration falls thereafter with time. In the intermediate region of molecular weights, the curves representing the changes in concentration with time are determined by a balance between the molecules that are degraded, and those formed due to degradation of molecules having a higher molecular weight. This diagram is thus the schematic representation of the kinetik degradation models that will be discussed later.

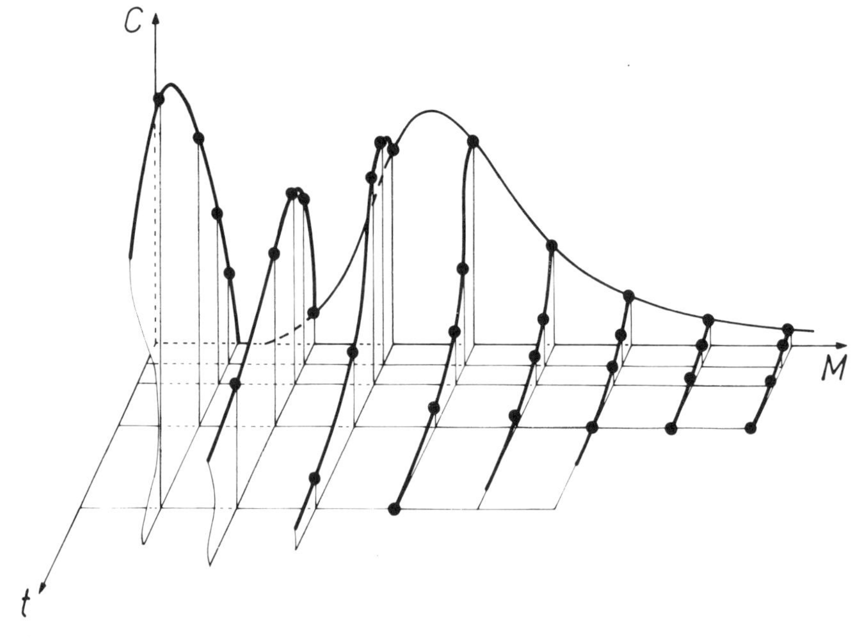

Fig. 18. Three-dimensional molecular weight (*M*)–concentration (*C*)–time (*t*) diagram representing kinetics of ultrasonic degradation

5.2. Evaluation of Molecular-Weight Averages

Most of the experimental investigations on degradation of polymers in solution due to ultrasound have been based on viscosity changes of the solution with the time of irradiation, or on the rate of production of free radical fragments. These techniques give only molecular-weight averages, and since the changes in molecular-weight distributions with time are not taken into account, the kinetic results are subjected to great uncertainties. This has been repeatedly pointed out by many investigators (*74, 75*). The fact that low molecular-weight components have only a very slight effect on the relative viscosity of the solution, led to the unusually high limiting molecular weights for degradation found by the earlier investigators. With the advent of gel permeation chromatography, it was demonstrated that the limiting molecular weights are considerably lower.

Schmid (*84*) was the first to carry out simplified kinetic calculations on the degradation of macromolecules. Starting from Eq. (13) he derived an equation relating the mean value of the molecular weight $\bar{M}_t$ of the polymer after the time of irradiation, to the limiting molecular weight M_e and the initial molecular weight M_i of the polymer sample:

$$\frac{M_e}{\bar{M}_t} + \ln\left(1 - \frac{M_e}{\bar{M}_t}\right) = -\frac{k}{c}\left(\frac{M_e}{M_0}\right)^2 \cdot t + \frac{M_e}{M_i} + \ln\left(1 - \frac{M_e}{M_i}\right) \tag{14}$$

where M_0 is the molecular weight of the monomer and c is the concentration of the polymer in moles per liter. In this equation the molecular-weight distributions are not taken into account, and the number- and weight-average molecular weights are not distinguished. Schmid's equation can be verified experimentally by plotting the value of $-\left[\frac{M_e}{\overline{M}_t} + \ln\left(1 - \frac{M_e}{\overline{M}_t}\right)\right]$ against the time of irradiation t. A straight line with the slope $\frac{k}{c}\left(\frac{M_e}{M_0}\right)^2$ should then be obtained. Schmid's experimental results confirmed this equation for polystyrene in toluene (Fig. 19). It is, however, somewhat surprising that in Fig. 19 straight lines are obtained, since Schmid committed the inconsistency of plotting viscosity-average molecular weights against the time of irradiation. In the derivation of Eq. (14) it is assumed that all molecular weights are expressed as number-averages. Most probably the polystyrene fractions used in this investigation had such broad molecular-weight distributions that the alteration of the factor $\overline{M}_w/\overline{M}_n$ was insignificant with time.

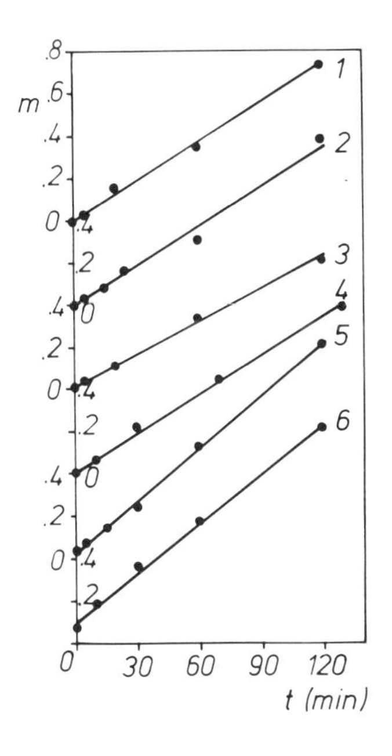

Fig. 19. Kinetic plot for degradation of polystyrene in toluene. Ordinate: $m = -[M_e/M_t + \ln(1 - M_e/\overline{M}_t)]$, abscissa: time of irradiation (t). Lines represent different initial molecular weights: *1:* 850000 *2:* 350000 *3:* 192000 *4:* 280000 *5:* 113000 *6:* 80000 M_e: 27000 [Ref. (*84*)]

Since the number of bonds broken in the polymer molecule is an unambiguous measure for degradation, the kinetics of the degradation reaction can be determined directly by estimating the free radicals formed. Most investigators used DPPH for this purpose. The first experiments were carried out by Henglein (*33, 34*) who demonstrated that the consumption of DPPH was proportional to the number of bonds broken in the polymer chain during irradiation. Chandra *et al.* (*12, 13*) investigated the degradation of butyl rubber in cyclohexane and toluene, comparing the kinetics of the reaction by viscosity measurements and free radical consumption using DPPH.

The general shape of the degradation rate curves was similar in all cases, but not identical. The rate given by estimation of DPPH was faster than that obtained from viscosity measurements, indicating that neither method is recommendable for the quantitative determination of the number of breaks occurring in the polymer molecule. The differences arise from the fact that the Mark-Houwink relation may not be strictly valid over the whole molecular-weight range investigated, and that the efficiency of DPPH as a free radical scavenger may not be unity as assumed in the calculations. Uncertainties in the trapping efficiency of DPPH have been pointed out by Thomas (*103*). This investigator analyzed the kinetics of degradation of polystyrene, polybutene and polymethylmethacrylate in benzene using DPPH to measure the free radical fragments formed during degradation. His results indicate that the rates of degradation are the same for the three polymers, but different for polylaurylmethacrylate and that the rate constants are proportional to the degree of polymerization (Fig. 20).

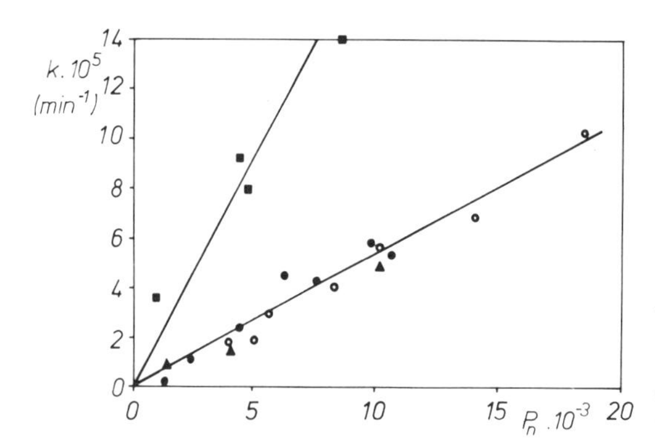

Fig. 20. Dependence of rate constants of degradation (k) on degree of polymerization (P_n) for different polymers. ▲: polymethylmethacrylate ●: polybutene ○: polystyrene ■: polylaurylmethacrylate [Ref. (*103*)]

In conclusion, the evaluation of molecular weight averages is an inefficient method for studying the kinetics of degradation of polymers in solutions by ultrasound. Except in those cases where monodisperse fractions are used, information about the dependence of the rate constants on the molecular weight of the polymer and the location of the point of breakage along the polymer chain is almost impossible to obtain. The determination of the molecular-weight averages of polymer fractions is only useful for qualitative, semiquantitative or comparative investigations of degradation.

5.3. Determination of Molecular-Weight Distributions

Although a great number of experiments has been carried out to study the kinetics of ultrasonic degradation of polymers in solution, only comparatively few investigators have analyzed the alteration of the molecular-weight distribution of the polymer during the course of degradation. The main reason for this is that only very few fractionation techniques are available, which have a resolution high enough for precise characterization of the degradation products. The earlier methods of fractional precipitation of the polymer using a solvent–nonsolvent system are very tedious, besides which, relatively large quantities of polymer are needed to obtain a great number of fractions, which in most cases are unobtainable. With the development of gel permeation chromatography, it became possible to obtain molecular-weight distributions with relatively good accuracy. Even though the molecular-weight distributions obtained from gel permeation chromatography are not absolute, because of the general lack of well-defined calibration standards and because of computational problems arising during the peak-broadening correction procedures, a comparison of any distributions obtained by the same method and under the same conditions shows quite accurately the differences existing between them. Since the reproducibility of gel permeation chromatography is very good and chromatograms are obtainable very quickly and using only few milligrams of substance, gel permeation chromatography is the best analytical method available today for studying the kinetics of ultrasonic degradation of polymers in solution.

The first analyses of degradation products using molecular-weight distributions have been performed by Jellinek and White (*39*). The molecular-weight distributions were obtained by fractionation of the dissolved polymer samples with a precipitant, followed by the viscosimetric characterization of the fractions thus obtained. The limiting degree of polymerization and the rate constants of degradation of polystyrene have been obtained in this way. Mostafa (*56*) also investigated the kinetics of degradation of polystyrene in benzene using the same method for the determination of the molecular-weight distributions. Using a similar technique, Schmid *et al.* (*91*) investigated the alteration of the inhomogeneity of the molecular-weight distributions of polymethylmethacrylate samples after different times of ultrasonic irradiation. Starting from relatively narrow polymer fractions, it was found that ultrasonic degradation first broadens the molecular-weight distribution, whereas at the later stages of degradation the distributions become narrower again. This is understandable, since polymer molecules, having a higher molecular weight, are more easily degraded. Gooberman and Lamb (*27*) investigated the degradation of polystyrene in benzene using turbidimetric titrations for the determination of the molecular-weight distributions of the degraded products. In his results the degraded polymer fractions reveal secondary peaks, indicating clearly that ultrasonic degradation is a *nonrandom* process.

Using the Baker-Williams fractionation method, Porter *et al.* (*75*) determined the molecular-weight distribution of ultrasonically degraded polyisobutene in n-hexadecane. Starting from very narrow polymer fractions as well as from fractions having a broad molecular-weight distribution, Porter *et al.* obtained in all cases two or three maxima in the molecular-weight distributions of the degraded products, thus giving further proof that ultrasonic degradation is nonrandom. The limiting degree of polymerization has been found to lie in the range from 100 to 200, *i.e.*, considerably lower than the value of 1000–2000 reported earlier by other investigators for several polymers.

With the aid of *gel permeation chromatography*, Porter *et al.* (*76*) evaluated the molecular-weight distributions of the degradation products of polyisobutene having an initial weight-average molecular weight of about 500 000 and an inhomogeneity of $\overline{M}_w/\overline{M}_n$ of approximately 2. With such broad molecular-weight distributions these researchers found that the inhomogeneity of the degraded products does not change appreciably with the time of irradiation, and that secondary peaks are not clearly shown. Investigations of Shaw and Rodriguez (*96*) on the degradation of polydimethylsiloxane having viscosity-average molecular weights from 2.4×10^5 to 1.3×10^6 showed that the molecular-weight distributions of the degraded products in toluene approach the same distribution after 120 min of irradiation at 36 watts. Smith and Temple (*97*) investigated more exactly the degradation of polystyrene fractions in tetrahydrofuran. Gel permeation chromatography was used to determine the molecular-weight distributions, the alterations of the latter being followed as a function of the time of irradiation. Typical results are depicted in Fig. 21, in which it can be seen that the change in the molecular-weight distribution of an initially

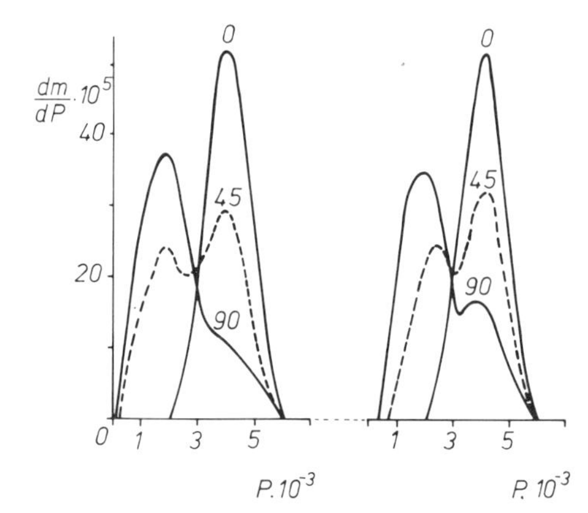

Fig. 21. Differential distribution functions of ultrasonically degraded polystyrene. *dm/dP* differential distribution function of degree of polymerization *dm* = mass of polymer having a degree of polymerization between *P* and *P* + *dP*; initial molecular weight $\overline{M}_w$ = 411 000, $\overline{M}_w/\overline{M}_n$ = 1.05; solvent: tetrahydrofuran; ultrasonic power of 48 watts at 20 kc/sec. Curves *left* are experimental; curves *right* are theoretical distribution functions calculated from simple degradation model. Parameter of curves is time of irradiation (in min) [Ref. (*97*)]

narrow polystyrene fraction shows a preferential breaking near the center of the molecule. Using a series of concentrations of polymers up to 20 g/l, the kinetics of degradation was analyzed by following the change in peak height of the molecular-weight distribution curves as a function of time. This procedure is allowed, since the heights of the distribution curves at a definite molecular-weight value are the weight fractions of that molecular species. For polymer fractions having either narrow or broad molecular-weight distributions, it was found that the rate of degradation was first order throughout the early stages in the reaction. This is clearly demonstrated in Fig. 22, where the change in the heights in the distribution curves of a broader polymer sample is plotted against the time of irradiation for different molecular weights. In addition to this, these investigators found that the variation of the rate constants was linear with the degree of polymerization. After 88 h of degradation at 20 kc/sec and a power of 48 watts, a limiting molecular weight of about 24000 was found, which is in fair agreement with the results of many investigators. More recent investigations on the degradation of polystyrene carried out at the author's laboratory indicate, however, that a value of 15 000 is more probable.

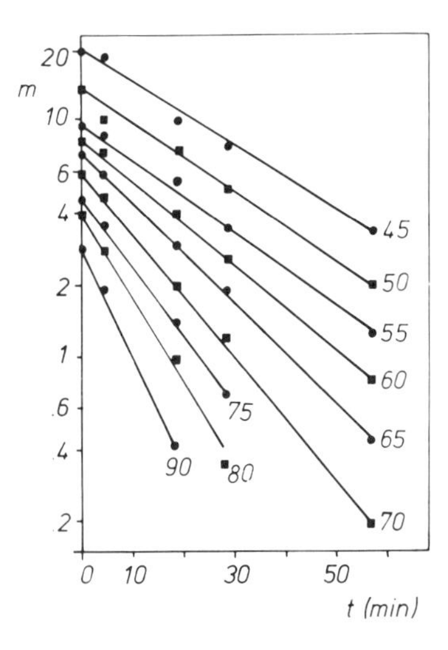

Fig. 22. First-order degradation rate plot for polystyrene ($\overline{M}_w$ = 433 200, $\overline{M}_w/\overline{M}_n$ = 1.71 in tetrahydrofuran at 48 watts and 20 kc/sec). Ordinate: weight fraction (m); abscissa: time of irradiation (t). Parameter of lines is degree of polymerization x 10^{-2} [Ref. (*97*)]

Smith and Temple (*97*) also carried out chemical degradation experiments of polystyrene, using benzoyl peroxide. The kinetics of degradation was first order in this case, too, but the rate constants were found to be independent of the molecular weight of the polymer. The molecular-weight distributions of the degraded products did not show secondary peaks, which was to be expected, since peroxide degradation is known to be a random process.

More detailed investigations were carried out by Basedow and Ebert (*2*) on the degradation of narrow dextran fractions at an ultrasonic frequency of 20 kc/sec in many solvents. The concentration of the polymer was 0.5%. The molecular-weight distributions were obtained to high accuracy by permeation on controlled pore glass. In all experiments it was found that rupture occurs near the center of the molecule. A typical set of distribution curves of degradation products is shown in Fig. 23. Similar results were found by Bradbury and O'Shea (*9*) for several proteins and Davison and Freifelder (*16*) for DNA.

The distribution curves shown in Fig. 23 have not been corrected for peak-broadening, since all techniques applied for this purpose introduced oscillations into the molecular-weight distribution curves, which cannot be tolerated in the investigation of degradation kinetics. Moreover, it has been verified that peak-broadening played a minor role in these investigations, since it was independent of the molecular weight of dextran at the conditions under which the chromatograms have been determined. A detailed discussion of the determination of the molecular-weight distributions of dextran on controlled pore glass, its precision and reproducibility has been given by Basedow *et al.* (*4*).

By selecting the sections in the range of the molecular-weight distribution curves where only degradation took place, the kinetics of degradation was followed by plotting the heights of the distribution curves at definite molecular weights against the time of degradation in a way similar to that of Smith and Temple (*97*). Straight lines were obtained in all cases, showing that the kinetics of the reaction was first order. The set of degradation lines corresponding to the distribution curves in Fig. 23

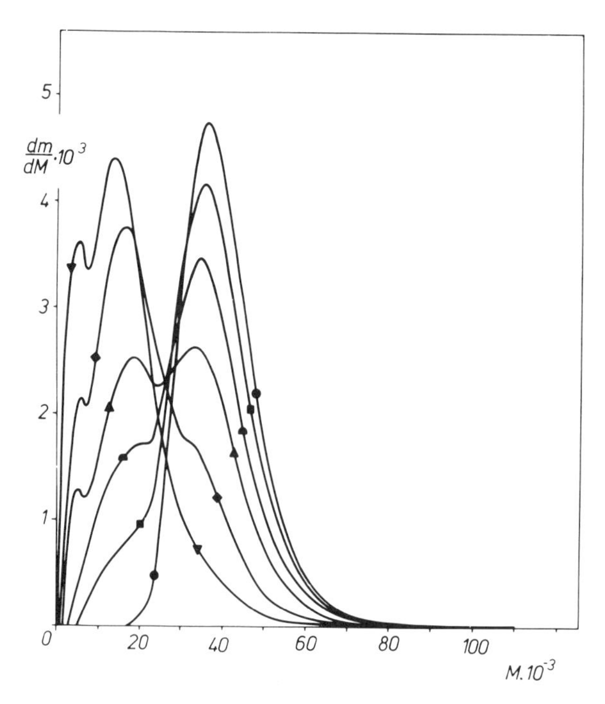

Fig. 23. Differential molecular-weight distribution curves of ultrasonically degraded dextran ($\overline{M}_w$ = 39700, $\overline{M}_w/\overline{M}_n$ = 1.05 in water at 24 watts/cm^2 and 20 kc/sec). Times of irradiation (min): ●: 0 ■: 10 ○: 20 ▲: 40 ◆: 80 ▲: 160 [Ref. (*2, 3*)] *dm*/*dM* = differential molecular weight distribution; dm = mass of polymer having a molecular weight between *M* and *M* + *dM*

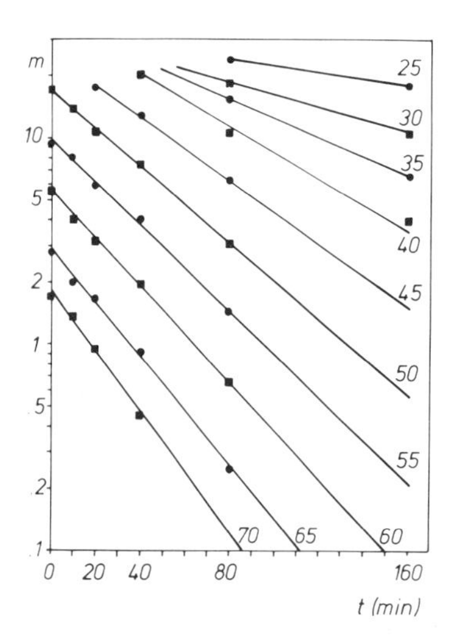

Fig. 24. First-order degradation rate plot of dectran (molecular-weight distribution curves from Fig. 23). Ordinate: weight fraction (*m*); abscissa: time of irradiation (*t*). Parameter of lines is molecular weight x 10^{-3} [Ref. (*2, 3*)]

is represented in Fig. 24. In dilute solutions the reaction order of ultrasonic degradation is found to be pseudo first order in most cases. The true reaction order is certainly more complex, since the rate constants decrease with increasing concentration in a nonlinear fashion (*24*). This is attributed by some investigators to entanglements and the building up of polymer networks, which slow down the degradation process.

From the slopes of the lines in Figure 24 the rate constants of degradation of dextran, having a definite molecular weight, can be calculated. A typical set of degradation constants is shown in Fig. 25, which demonstrates that the rate constants

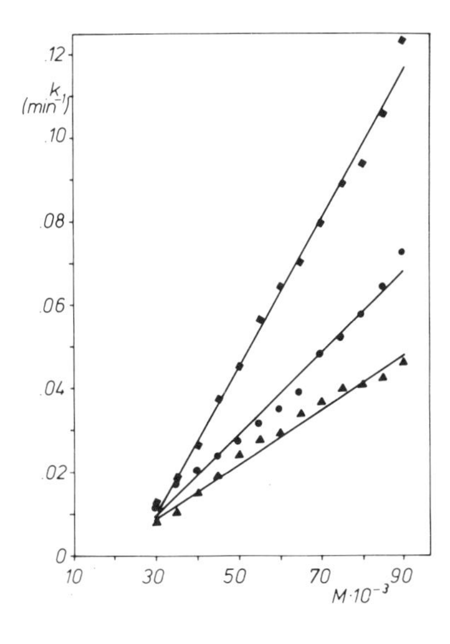

Fig. 25. Dependence of degradation rate constants (k) on molecular weight (M) of dextran in different solvents. Ultrasonic intensity 24 watts/cm^2 at 20 kc/sec. Solvents are: ▲: water ●: solution of 10% $MgSO_4$ in water ◆: formamide [Ref. (*2*)]

increase linearly with increasing molecular weight. An extrapolation of the lines, which represent the dependence of the rate constants on the molecular weight, to the rate constant zero should give in principle the *limiting molecular weight,* below which degradation ceases. For dextran the values thus obtained for several different solvents are of the order of 20000. Long-time investigations of degradation, however, showed that the limiting molecular weight of dextran in water was only approxi-

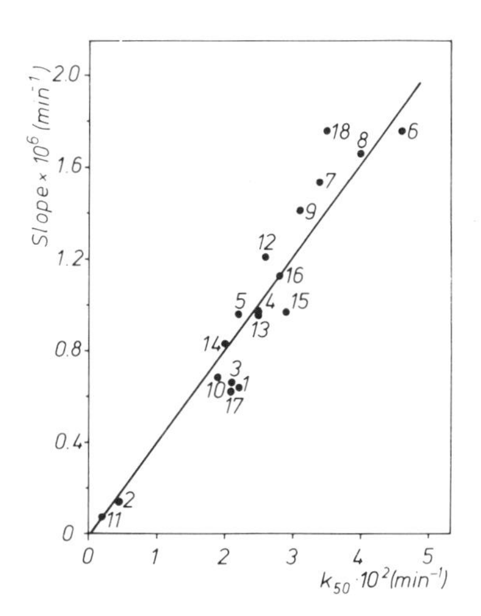

Fig. 26. Plot of slopes of straight lines, which give dependence of degradation rate constants on molecular weights (Fig. 25), versus degradation constant of dextran with molecular weight of 50.000 (k_{50}). Numbers refer to different solvents: *1:* water (ultrasonic intensity 24 watts/cm^2) *2:* water (ultrasonic intensity 5 watts/cm^2) *3:* water (different initial molecular-weight distribution) *4:* deuterium oxide *5:* dimethylsulfoxide *6:* formamide *7:* ethylene glycol *8:* glycerin *9:* ethanolamine *10:* 20% methanol in water *11:* water saturated with diethyl ether *12:* water with 0.1% potassium palmitate *13:* 1% $MgSO_4$ in water (20 °C) *14:* 1% $MgSO_4$ in water (50 °C) *15:* 10% $MgSO_4$ in water *16:* 10% glucose in water *17:* 10% urea in water *18:* 10% urea in dimethylsulfoxide. If not indicated, ultrasonic intensity was 24 watts/cm^2 at 20 kc/sec [Ref. (*2, 3*)]

mately 7000, at an ultrasonic intensity of 24 watts/cm^2. This proves that the degradation kinetics are altered at very low molecular weights.

An interesting feature is the fact that the slopes of the curves in Fig. 25, which represent the dependence of the rate constants on the molecular weight, are proportional to the rate constants themselves. In Fig. 26 the slope is plotted against the rate konstant k_{50} of a dextran having a molecular weight of 50000. A straight line, which runs through the origin of the coordinates, is obtained. This line is independent of the intensity of the ultrasonic waves, the intensity of cavitation, the solvent and other experimental conditions. It depends only on the nature of the polymer investigated. Conclusions drawn from this fact will be discussed later in connection with the mechanisms of degradation.

5.4. Degradation Models

The aim of studying the kinetics of ultrasonic degradation is to develop a kinetic model for the prediction of the molecular-weight distributions of the degraded polymer fractions after definite periods of ultrasonic irradiation. In order to establish a degradation model, many factors affecting the kinetics of the reaction must be known, such as the variation of the rate constants with the molecular weight, the order of the reaction, the location of the point of breakage and the limiting molecular weight. A rigorous treatment of polymer degradation requires the simultaneous solution of a large number of complex differential rate equations, one for each molecular-weight species present in the sample. In addition to this, the rate constants entering into these equations must be known. Since these parameters must be varied over wide ranges, most kinetic models are based on iterative methods for finding those values of the variables that give the best agreement with experiments. Analytical solutions are obtainable in very few cases only, so that commonly numerical techniques, which afford the use of large computers, provide the only source of precise investigation of the degradation kinetics.

In principle the problem is to find a numerical expression for the molecular-weight distribution curves in Fig. 23 as a function of time. Heymach and Jost (*36*) developed a general computer program that can incorporate any type of degradation kinetics and thus provide a suitable basis for comparing the various models. In order to carry out the computations, the chain is divided into S segments each having n_s monomer units per segment. The inputs to the program include a rate equation for degradation, a distribution function of the fragments arising from the degradation of a molecule and an initial molecular-weight distribution. The solution proceeds by calculating the loss of molecules from the rate equation in differential form, for the first time increment, starting at the lowest value of S, such that $S \cdot n_s > P_e$, where P_e is the limiting degree of polymerization. The loss of molecules is then doubled, since each degraded molecule yields two smaller ones, and is redistributed at lower values of S according to the function used for this purpose. This step is repeated for the first time increment with the next larger S, and so on, until the entire molecular-weight distribution has been traversed. The entire computation cycle is repeated for the next time increment, and so on, until the desired degradation time has been reached. Heymach and Jost compared the degradation models proposed by many investigators and found considerable differences between them, but on account of the lack of precise experimental data, it was impossible to select the best model.

Glynn and van der Hoff (*23*) also developed a degradation model which gives the molecular-weight distributions of the degraded polymer after a certain number of chain ruptures. In this model degradation is described in terms of the probability that a molecule of a given length will break, and the probability that a molecule of a definite length will result from this rupture. Any form of both these probability distribution functions and any initial molecular-weight distribution of the polymer can be used in this model. A limiting degree of polymerization can also be included. In its more general form the probability of choosing a molecule for rupture is taken to be proportional to its molecular weight raised to a power b, and the localization of the point of breakage along the chain is represented by a Gaussian distribution about the midpoint of the extended molecule with standard deviation $\sigma = r \cdot P$ and truncated at the chain ends, where P is the degree of polymerization of the initial polymer molecule and r is a constant. According to the values of b and r some limiting cases shall be considered. For $b = 0$ the probability of choosing a molecule for rupture is proportional to its number fraction, whereas for $b = 1$ it is proportional to its weight fraction; if r is very large, the probability of rupture along the polymer chain is constant, *i.e.*, random breakage occurs, whereas for very small values of r, breakage of the molecules will take place at their midpoint only, *i.e.*, center breakage will occur. The calculations were carried out in terms of a degradation index DI, defined as:

$$DI = \frac{\overline{M}_n(o)}{\overline{M}_n(t)} - 1 \tag{15}$$

where $\overline{M}_n(o)$ is the number-average molecular weight of the original sample and $\overline{M}_n(t)$ that of the degraded sample after t breaks of the molecule. These investigators tested their degradation model using polystyrene samples with narrow molecular-weight distributions, which were degraded in tetrahydrofurane under variable experimental conditions (*22–24*). The experimental molecular-weight distributions were compared with those calculated for several values of the adjustable parameters b and r of the kinetic model. It was found that the experimental results disagreed with the calculated molecular weight distributions, if random or center breakage was assumed. The calculated molecular-weight distributions closely followed the experimentally obtained distribution curves for values of b from 1.0 to 1.25 and for values of r from 0.15 to 0.35. As an example of their results, the degradation of polystyrene having a weight-average molecular weight of 860000 and $\overline{M}_w/\overline{M}_n = 1.15$ in tetrahydrofuran is shown in Fig. 27 and 28. In these investigations the experimental molecular-weight distribution curves were obtained by gel permeation chromatography and corrected for instrumental spreading. The number-average molecular weight of the polymer fractions, which must be known to calculate the degradation index, was calculated from the corrected chromatograms. Moreover, it was demonstrated that the experimental conditions of concentration, temperature and ultrasonic intensity had no effect on the course of the changes in the molecular-weight distributions, although the rate of degradation was greatly affected.

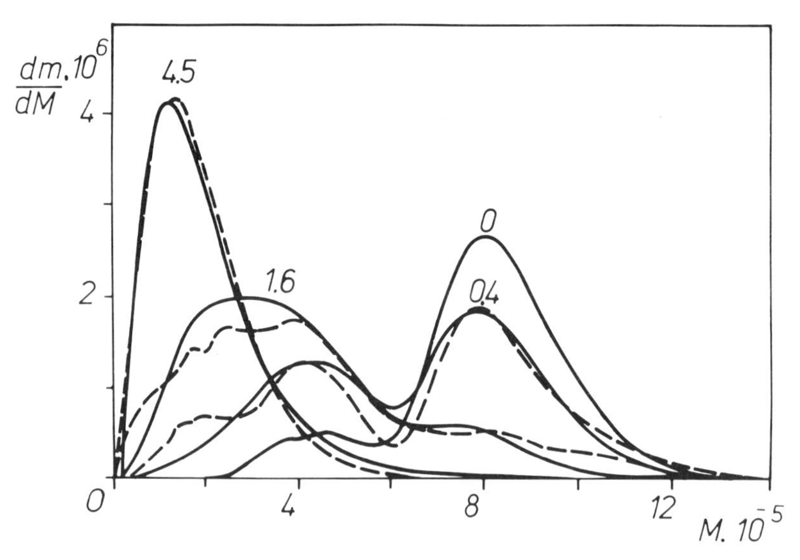

Fig. 27. Comparison of experimental molecular-weight distributions (broken lines) with distributions calculated from degradation model (solid lines r = 0.15; b = 1). Polystyrene ($\bar{M}_w$ = 860000, $\bar{M}_w/\bar{M}_n$ = 1.15) at 35 watts/cm^2 and 22 kc/sec in tetrahydrofuran. Parameter of curves is degradation index DI [Ref. (*22*)].
dm/dM = differential molecular weight distribution; dm = mass of polymer having a molecular weight between M and $M + dM$

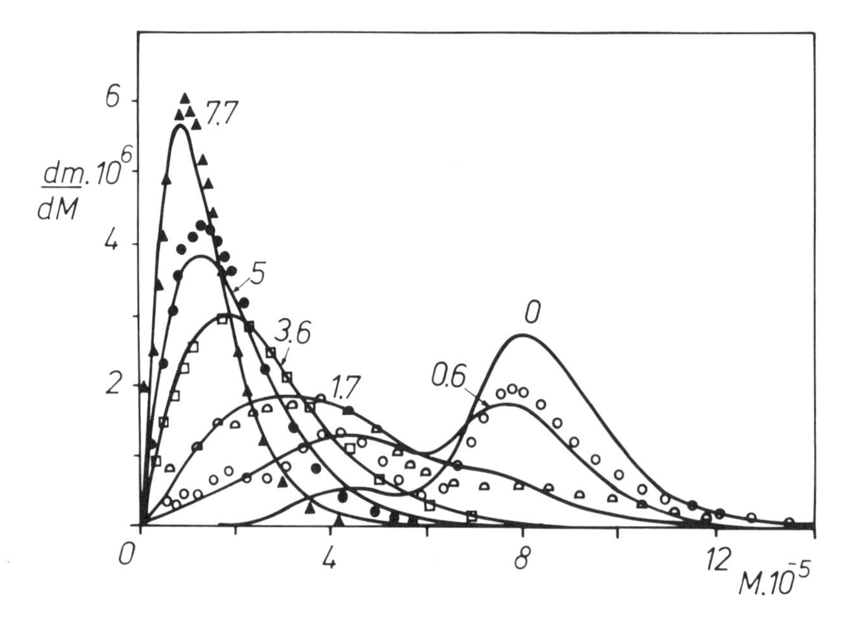

Fig. 28. Comparison of experimental molecular-weight distributions (symbols) with distributions calculated from degradation model (lines r = 0.35; b = 1.25). Polystyrene ($\bar{M}_w$ = 860000, $\bar{M}_w/\bar{M}_n$ = 1.15) at 28 watts/cm^2 and 22 kc/sec in tetrahydrofuran. Parameter of curves is degradation index DI [Ref. (*23*)]

Suppanz and Ebert (*100*) used a different method for calculating the molecular-weight distributions of the degradation products of dextran fractions after different periods of time. Starting from the initial molecular-weight distribution of the polymer sample, the calculation algorithm gives the numerical solution of the rate equation for each degree of polymerization, without introducing any simplifications. The kinetic model is based on the rate equation:

$$-\frac{dC(M_i)}{dt} = C(M_i) \cdot K \cdot M_i^{\beta} \tag{16}$$

where $C(M_i)$ is the concentration of molecules with molecular weight M_i, K is a constant which depends only on the experimental conditions of the degradation experiment and β is an exponent, which is choosen by trial and error and which represents the dependence of the individual rate constant on the molecular weight of the polymer molecule considered. Several reaction orders were tried, and it was found that first-order kinetics gave the best agreement between the calculated and the experimental molecular-weight distributions. The location of the point of breakage along the polymer chain was given as an integral quotient of the whole polymer chain, and was kept constant for each computation. This quotient restricted the model in some ways, but assured exact solutions of the rate equations. Several quotients were analyzed, detailed results being computed for quotients of 1/1, 1/2, 1/3, 1/9 and for the scission of a small fragment of constant length from the polymer chain. The first case, where the resulting fragments have equal size, is the center breakage model.

The computations were carried out using the experimental values of the rate constants, as calculated from the experimental degradation curves in the range where only degradation occurred (*3*). Some slight variations of the rate constants was allowed during the computations. Some typical results for the degradation of dextran in water are shown in Figs. 29 and 30. From these figures it becomes clear that the center breakage model gives the best agreement between the experimental and the calculated molecular-weight distributions, and that the best value of β lies between 1.0 and 1.5. Once the ratio of lengths of the degradation fragments has been determined and the value of β is known, the kinetics of degradation depends only on the value of K and the initial molecular-weight distribution of the polymer sample. Since, according to Eq. (16), the rate constant $k(M_i)$ of degradation of molecules with a molecular weight M_i is given by:

$$k(M_i) = K \cdot M_i^{\beta} \tag{17}$$

the coefficient K includes all the parameters which affect ultrasonic degradation, such as intensity and frequency of ultrasound, intensity of cavitation, temperature, solvent and the nature of the investigated polymer. Although the algorithm of this model gives the exact solution of the rate equations, this procedure has limited application, since it does not permit deviations of the ratio of lengths of the degradation fragments from integers.

From the observations of this chapter, it can be safely concluded that in dilute solutions ultrasonic degradation follows a *first-order* reaction, the rate constants of

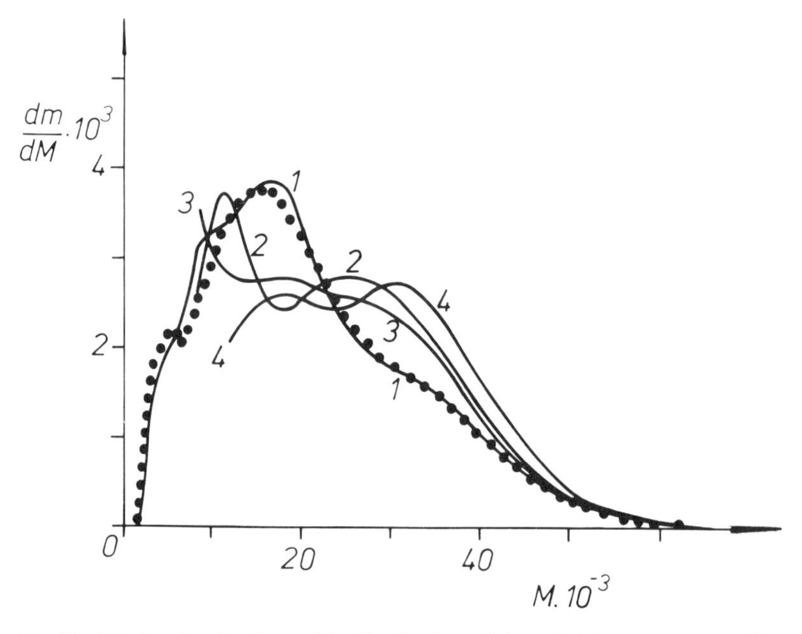

Fig. 29. Calculated molecular-weight distributions of degraded dextran in water, for $\beta = 1$ and assuming variable fragmentation ratios. Initial molecular weight $\bar{M}_w = 39700$, $\bar{M}_w/\bar{M}_n = 1.05$; degradation time 80 min at 24 watts/cm^2 and 20 kc/sec. Numbers indicate different fragmentation ratios: *1* = 1:1 *2* = 2:1 *3* = 3:1 *4* = 9:1 ●●●●: experimental curve [Ref. (*100*)]

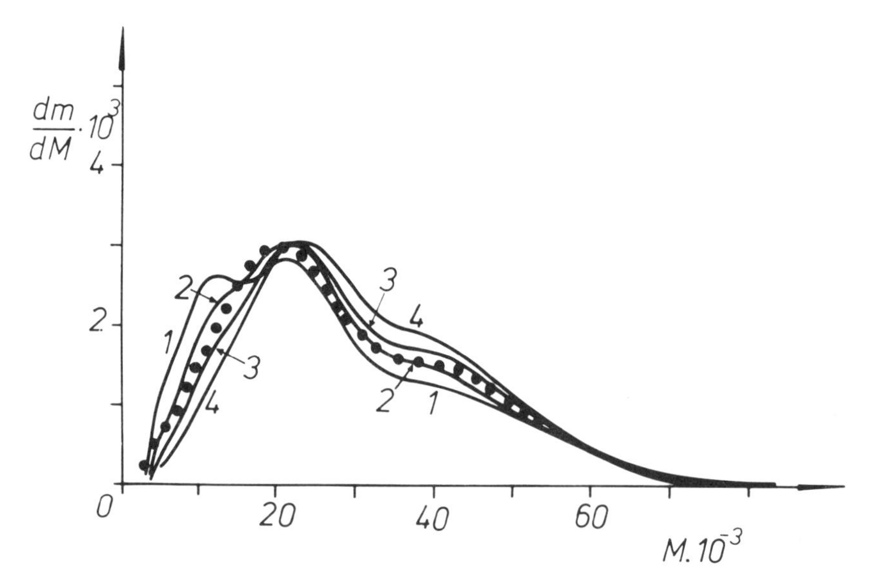

Fig. 30. Calculated molecular-weight distributions of degraded dextran in water, for fragmentation ratio of 1:1 and assuming variable values of β. Initial molecular weight $\bar{M}_w/ = 53200$, $\bar{M}_w/\bar{M}_n = 1.08$; degradation time 40 min at 24 watts/cm^2 and 20 kc/sec. Numbers indicate different values of β: *1:* $\beta = 0.5$ *2:* $\beta = 1$ *3:* $\beta = 1.5$ *4:* $\beta = 2.5$ ●●●●: experimental curve [Ref. (*100*)]

degradation being *proportional* to the molecular weight of the polymer molecule, or possibly to the molecular weight raised to a power of 1.25. Finally it is shown that the molecules break *preferentially* at points close to their *midpoints,* although exact center breakage seems not to take place.

6. Mechanisms of Ultrasonic Degradation

6.1. Fundamentals of Mechanochemical Reactions

In chemical reactions initiated or accelerated by mechanical action, the estimation of the elastic energy involved is extremely difficult. This is mainly because of the lack of clear concepts concerning the mechanism of these reactions, and because of the methodological difficulties in measuring the amount of mechanical energy absorbed as well as the magnitude of the stresses that actually operate. Mechanochemical reactions are complex multistage processes, which include mechanical deformation of the molecule before the actual chemical reactions take place. In the kinetics of mechanochemical reactions we may consider reactions in constant mechanical fields and reactions in variable mechanical fields.

In constant mechanical fields elastic energy is absorbed rapidly and the rate of the mechanochemical process is determined by the rate of the chemical reaction. An example of this type of reaction is the degradation of polymers on extension of polymeric films or fibers by the application of a constant stress. In variable mechanical fields the rate of absorption of elastic energy is much slower than the rate of the actual reaction, which follows as a result of the absorption of this energy. An example of this type of reaction is the whole complex of chemical processes that take place in polymer solutions under the influence of ultrasonic irradiation. In variable mechanical fields the efficiency of the mechanochemical process depends on the frequency of mechanical action and increases with the intensity of the latter. The rate-limiting stage of such processes is the redistribution of elastic energy in the molecule, *i.e.,* the rate of the entire reaction and the temperature variation of rate constants depend on the physicomechanical properties of the substance. In the case of macromolecules the absorption of elastic energy and the rate of the mechanochemical reaction depend on the length of the chain, *i.e.,* on the degree of polymerization. Other conditions being constant, the kinetics of degradation of polymers is determined by the variation of the elementary rate constant with the length of the molecule.

Mechanochemical reactions can occur with a comparatively low average level of elastic energy per unit volume of substance. This is generally not observed in other reactions initiated by physical methods, *e.g.,* photochemical reactions. According to Butyagin *(11)* there are two direct mechanisms for the conversion of elastic into chemical energy: one is associated with change in the interatomic distances, *i.e.,* with deformation of electron clouds, and the other is associated with vibrational excitation of the bonds during the dissipation of elastic energy as heat. The first mechanism is an equilibrium process, since the equilibrium Maxwell-Boltzmann distribution of energy among the degrees of freedom persists in the system; the second is a nonequilibrium process.

In the equilibrium process chemical bonds are deformed. Stretching a bond decreases its energy; at equilibrium, the deformation energy E_{def} is:

$$E_{def} = \int_{r_0}^{r} F \cdot dr \tag{18}$$

where F is the force acting on the bond, r_0 and r are the length of the initial and the deformed bond respectively. The deformation energy can be calculated assuming a Morse potential function. The activation energy for decomposition E depends on the force F according to the relation (*11*):

$$\frac{E}{E_0} = \sqrt{1 - \frac{F}{F_0}} - \frac{F}{2F_0} \ln \left(\frac{1 + \sqrt{1 - \frac{F}{F_0}}}{1 - \sqrt{1 - \frac{F}{F_0}}} \right) \tag{19}$$

where F_0 is the tensile strength of the bond and E_0 the bond energy. This equation describes the simplest case of the conversion of mechanical into chemical energy. For a macromolecule with chain length equivalent to 10, 20, 50, 100 and ∞ bonds, the function $E(F)$ is represented in Fig. 31.

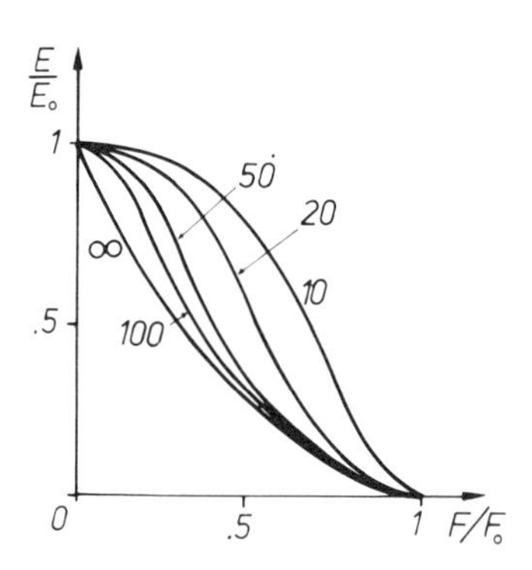

Fig. 31. Plot of function E/E_0 versus F/F_0 for different numbers of deformed bonds in chain. Parameter of curves is number of deformed bonds [Ref. (*11*)]

When the bonds of a molecule are mechanically strained, the decomposition rate constant k_M may be given by an expression similar to the Arrhenius equation in which the activation energy for thermal decomposition is replaced by the energy of an elastically deformed bond $E(F)$:

$$k_M = k_{O,M} \cdot \exp\left[\frac{-E(F)}{RT} \right] \tag{20}$$

where $k_{O,M}$ is the pre-exponential factor. From this analogy, it can be seen that in an equilibrium process the mechanical stress decreases the energy barrier of the thermal reaction. The mechanical dissociation of bonds without thermal fluctuations is possible only if the stress exceeds the bond tensile strength F_0, and is applied for a period of time less than the vibration period of the bond, *i.e.*, of the order of 10^{-13} sec. Since the bond lengths are of the order of 10^{-8} cm, the velocity of propagation of stress must be greater than 10^5 cm/sec, *i.e.*, comparable to the velocity of sound. This type of purely mechanical rupture occurs under the action of shock waves. The decomposition of strained bonds of polymer chains by mechanical action is confirmed by the formation of free radicals at the moment of dissociation.

In the nonequilibrium mechanism the rate of the mechanochemical reaction is determined by the rate of relaxation of the elastic energy in the system. The reaction takes place during the removal of stress or during its redistribution at the chemical bonds. When the stress is removed, the entire flux of elastic energy is liberated as heat at the bonds; therefore, for a certain period of time τ^*, these bonds are in a vibrationally excited state. During this time, the excitation energy can either pass to other degrees of freedom of the system, or it can produce the mechanochemical reaction. If τ_{chem} is the characteristic reaction time, the ratio τ^*/τ_{chem} gives the probability of the reaction taking place during the lifetime of the vibrationally excited state. If $\tau^*/\tau_{chem} \geqslant 1$, the excitation of the bonds always leads to a reaction. According to Butyagin (*11*) this inequality holds if:

$$\Delta E^*_{vibr} \geqslant \frac{E(F)}{\ln(\tau^*/\tau_0)} \tag{21}$$

where ΔE^*_{vibr} is the critical vibrational energy necessary for the dissociation of the bond within the period of time τ^* and τ_0 is the vibration period of the atoms forming the bond. In polymers the dissociation of vibrationally excited bonds produces hot macroradicals capable of further chemical reactions, *e.g.*, the decomposition into products of low molecular weight.

In ultrasonic chemistry and in the chemistry of shock waves other mechanisms, which explain the mechanochemical reaction as a result of the generation of local regions, where the elastic energy is concentrated before being absorbed, are used. In high-intensity ultrasonics in liquids, the local absorption of elastic energy is closely related to the phenomenon of cavitation. In depolymerization reactions free radicals are usually formed. In this case the shear stresses not only dissociate the bonds in the main chain, but also separate the radicals formed, preventing in this way their recombination. An extensive review of the experimental work on the mechanical degradation of polymers has been given by Porter and Casale (*77*).

6.2. Force Necessary to Rupture Covalent Bonds

In order to calculate the force necessary to rupture a covalent bond, the potential curve of the two atoms involved must be known. The relation proposed by Morse, which expresses the potential energy E as a function of the potential energy E_0 at the minimum of the potential curve and the distance r between the two atoms, can be used for this purpose (*6, 98, 99*):

$$E = E_0 \cdot e^{-2a(r-r_0)} - 2E_0 \cdot e^{-a(r-r_0)} \tag{22}$$

in which r_0 is the equilibrium distance and a is the so-called Morse-constant given by the formula:

$$a = \sqrt{\frac{8\pi^2 c\bar{m} x\nu_0}{h}} \tag{23}$$

In this formula c is the velocity of light, $\bar{m}$ is the reduced mass, ν_0 is the specific frequency of the bond, x is the anharmonic factor and h is Planck's constant. The anharmonic factor is given by:

$$x = \frac{h\nu_0}{4E_0} \tag{24}$$

The force F between the two atoms can be calculated from Eq. (22), *i.e.*:

$$F = \frac{\partial E}{\partial r} = -2a E_0 \cdot e^{-2a(r-r_0)} + 2a E_0 \cdot e^{-a(r-r_0)} \tag{25}$$

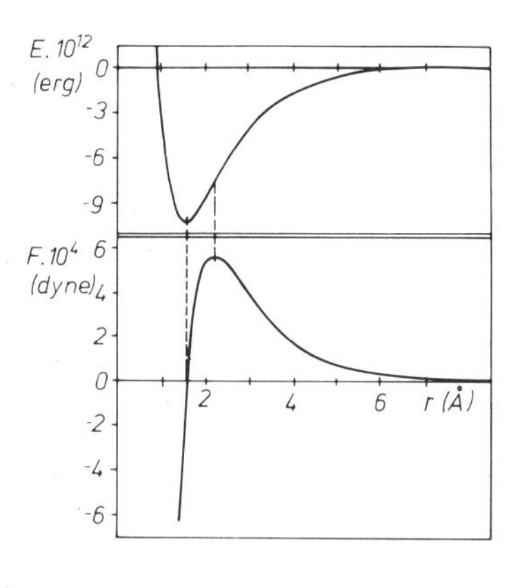

Fig. 32. Potential energy (E) and force (F) between two aliphatic bound carbon atoms as function of distance (r) between them [Ref. (*6*)]

The curves of E and F are represented in Fig. 32. The force F has a maximum value at a certain distance r_{max} given by:

$$\frac{\partial F}{\partial r} = 0 = 4a^2 E \cdot e^{-2a(r_{max}-r_0)} - 2a^2 E_0 \cdot e^{-a(r_{max}-r_0)} \tag{26}$$

and

$$r_{max} = \frac{ar_0 + \ln 2}{a} \tag{27}$$

If a C–C bond is considered, we have: $m = 9.95\text{x}10^{-24}$ g, $\widetilde{\nu}_0 = 810$ cm^{-1}, $\nu_0 = c \cdot \widetilde{\nu}_0 = 2.428\text{x}10^{13}$ sec^{-1}, $x = 3.6\text{x}10^{-3}$, $a = 1.025\text{x}10^8$ cm^{-1}, $r_0 = 1.55\text{x}10^{-8}$cm and $E_0 = 1.11\text{x}10^{-11}$ erg. Inserting these values into Eqs. (*27*) and (*25*) we get: $r_{max} = 2.23\text{x}10^{-8}$ cm and $(F_{C-C})_{max} = a \cdot E_0/2 = 5.64\text{x}10^{-4}$ dynes.

In a similar way it is possible to calculate the tensile strength of any chemical bond. For most polymers the calculated values for this force lie between $5-15\text{x}10^{-4}$ dynes. According to Lippincott and Schröder (*47*) the following empirical function gives results which are in better agreement with the data obtained from infrared spectra:

$$E = E_0 - E_0 \cdot \exp\left[\frac{-n(r-r_0)^2}{2r}\right] \qquad (28)$$

where n is a factor determined by the atoms forming the bond, given by:

$$n = \frac{k_0 \, r_0}{E_0} \qquad (29)$$

In this equation k_0 is the force constant of the bond at zero displacement of the nuclei. Using the same procedure described above, Davison and Levinthal (*15*) calculated the critical tensile strength of the C–C bond as $(F_{C-C})_{max} = 8.1\text{x}10^{-4}$ dynes, which is in reasonable agreement with the value calculated using the Morse function. The exact calculation of the maximum tensile strength of a covalent bond in polymer chains is difficult, due to the lack of precise information about the parameters of the bond involved, as well as uncertainties in the exact shape of the potential function of the bond. Fortunately, the calculation of the maximum tensile strength is relatively insensitive to the shape of the potential curve and depends primarily on the bond distance and the maximum energy of the bond, *i.e.*, of those quantities, which are most accurately known. For that reason, within a factor of two, the calculations above certainly represent the real conditions of bond strengths in macromolecules.

6.3. Direct Action of Ultrasonic Waves on Macromolecules

In order to explain the depolymerizing action of ultrasound several mechanisms have been proposed, most of them based on a direct action of ultrasound on the polymer molecules, none of which, however, was able to explain all the effects observed. Various attempts have been made to develop mathematical theories for degradation, taking into account most of the data available. It has been possible to reach fair agreement with experimental results, but in view of the incorrect assumptions made and the many arbitrary parameters used in the theories, the conclusions are not very convincing. Reviews of the older theories concerning degradation were given by Wilke and Altenburg (*107, 108*), Mark (*49*) and Él'Piner (*17*).

Since it has been found that chemical effects are not responsible for degradation and a limiting chain length has been detected, Schmid and Rommel (*85*) suggested that degradation was mechanical. According to Schmid (*86*) the vibrations of the solvent molecules in the ultrasonic field are different from those of the polymer chain; this gives rise to friction forces between the macromolecule and the solvent, which are able to break chemical bonds. For his experimental conditions, Schmid calculated the velocity amplitude of the vibrating solvent molecules, and obtained values of the order of 40 cm/sec. As the macromolecules have a different mass than the molecules of the solvent, inertial forces arise when the particles are vibrating under the action of ultrasonic waves. In order to investigate this effect, Schmid and Beuttenmüller (*87*) degraded polystyrene at a frequency of 284 kc/sec and an intensity of 50 watts/cm^2 in mixtures of toluene and carbon tetrachloride having different densities. The results have already been given in Section 4.2. (Fig. 16), the dashed line at $\rho = 1.058$ g/cm^3 representing the condition where the densities of the mixed solvent and the polymer are equal. If the difference in *inertia* of the macromolecules and the solvent molecules were responsible for degradation, deviations should be observable at the point of equal density. Since no deviation has been found, it may be concluded that inertial forces are not the cause of degradation.

Schmid (*86*) then analyzed the possibility of the macromolecules being broken by *frictional* forces. In a first case he considered the polymer molecule being represented by a frictionless thread, having spheres of radius r placed at regular intervals along its axis and rigidly held at one end in the solution. The total friction force f may be calculated by Stoke's law:

$$f = P_n \cdot 6\pi\, \eta\, r\, U_0 \tag{30}$$

where P_n is the degree of polymerization, η is the viscosity of the solvent and U_0 is the maximum velocity of the solvent molecules. Jellinek and White (*40*) computed the frictional force for polystyrene in benzene for certain experimental conditions ($U_0 = 51.1$ cm/sec, $r \approx$ radius of the benzene ring = $3\text{x}10^{-8}$ cm, viscosity of 0.0062 dynes. sec./cm^2, $P_n = 3000$) and obtained a value of $5.37\text{x}10^{-4}$ dynes. According to the calculations described in Section 6.2 this force is in the right order of magnitude for rupture. No evidence, however, is given for a polymer molecule being rigidly fixed at one end.

In a second model Schmid (*86*) considered that the polymer molecule moved freely within the solution. Under this condition the macromolecule will move with the solvent to an extent depending on the ratio of the inertia of the chain to the frictional force acting on it. Schmid, and Jellinek and White showed that in this case the frictional force f is given by:

$$f = \frac{2\, P_n\, \pi\, M_0\, \nu\, U_0}{N} \tag{31}$$

where M_0 is the molecular weight of the monomer, N is Avogadro's number and ν is the frequency of the ultrasound. Calculations based on this equation show that for the same experimental conditions (frequency of 500 kc/sec) as in the first case, the force acting on the polymer molecule is only $8.31\text{x}10^{-11}$ dynes, which is obviously insufficient to break a C–C bond. From these considerations Schmid derived the wrong conclusion that macromolecules in solution degrade only if they are more or less entangled, because only then friction forces are in the right order of magnitude.

Jellinek and White (*40*) calculated the *impact* forces set up on a polymer molecule fixed at both ends, during the collision with solvent molecules due to ultrasonic irradiation. Considering that the time of impact is short, so that the polymer chain may be considered rigid, the following equation was evaluated for the force:

$$f = \frac{4 P_n \rho l r c U_0}{\pi} \tag{32}$$

where c is the velocity of sound, ρ is the density of the solvent and l the distance between the monomer units. For the same experimental data used above, with l = 3×10^{-8} cm, an impact force of 1.88×10^{-5} dynes results, which again is too small to break a C–C bond. In order to elucidate the mechanism of ultrasonic degradation Mostafa (*57–60*) took all experimental results concerning the dependence of the degradation rate and limiting degree of polymerization on ultrasonic frequency and intensity and on the initial chain length of the polymer into consideration. He assumed the polymer to be a frictionless thread, having spheres, representing the monomers, attached to its axis at equal distances. He further assumed that the chain is fixed at both ends and orientated at right angles to the direction of propagation of the ultrasonic waves. With the same arguments as Schmid's he concluded that the motion of solvent molecules across the monomer units will give rise to frictional forces. Assuming that oscillations of the polymer chains are possible, resonance will take place at a definite frequency for a given chain length. Considering the fundamental mode of vibration of the macromolecule Mostafa (*60*) calculated the kinetic energy of oscillation of the chain. With the condition that *entanglements* and aggregation of the polymer molecules are present, the calculated kinetic energy of oscillation for polystyrene with a degree of polymerization ranging from 100 to 1000 at an ultrasonic frequency of 1000 kc/sec is of the order of 2×10^{-10} erg. Since the energy required to break a C–C bond is 1.1×10^{-11} erg (*6*), the above considerations show that long chain molecules can be ruptured by ultrasonic waves alone, provided the ultrasonic intensity is high enough.

All the mechanisms described in this chapter require the presence of entanglements in order to explain degradation. No degradation should therefore occur in extremely dilute solutions, where the macromolecules can move freely, and will be swept away following the ultrasonic oscillations, whatever the length of the chain. At very high concentrations, on the other hand, entanglements will be excessive and the polymer molecules will form a highly constrained network, where the flow of solvent molecules is greatly reduced. According to Schmid (*86*) the degradation should cease altogether in gels.

In practice, however, the polymer molecules are not exactly fixed at both ends, nor are they oriented specifically at right angles to the direction of propagation of the ultrasonic waves. Furthermore, macromolecules in solution are not stretched out in straight segments, but are moderately curled forming more or less statistical coils. In addition to this, degradation is readily possible in very dilute solutions, as well as in gels. For example, native dextran gels having viscosities of the order of

10^4 dynes x sec/cm^2 and dextrans crosslinked with epichlorohydrin (Sephadex), which in the presence of water forms rigid gels, are easily degraded by ultrasound.

From these experimental facts it follows that the models described above cannot explain the observed phenomena. They can at most explain degradation at rather definite concentrations and under very *limited* experimental conditions, *e.g.*, those imposed by Mostafa (*59*). Nevertheless, their discussion is useful to the understanding of the manifold actions of ultrasound on polymer solutions.

6.4. Shear Degradation

Degradation of polymers in solution by the action of ultrasound evidently occurs as a result of the stresses set up in the molecule by hydrodynamic shear. A quantitative treatment of the problem is complex because the process is usually accompanied by turbulence, and in the case of very high-frequency ultrasound further complications arise by intense local adiabatic heating of the liquid due to cavitation. Since no adequate theory of non-Newtonian flow under conditions of high shear has been developed up to now, the mathematical treatment of the problem is restricted to highly simplified and somewhat idealized models. Although all calculations of shear stresses are necessarily approximated, a comparison of the calculated force required to cause dissociation of the molecule with the theoretical value predicted from the potential function can provide valuable information about the elementary chain scission mechanism. For this reason a theoretical treatment of shear degradation is essential for the understanding of ultrasonic depolymerization, because also in this case, shear forces are the ultimate factors which cause the scission of the molecules.

A variety of shearing devices has been used to investigate the degradation of macromolecules. Davison and Levinthal (*15*) carried out degradation experiments in capillaries, maintaining conditions of laminar flow. Assuming that the frictional drag per unit length f is constant along the molecule, they proposed the following

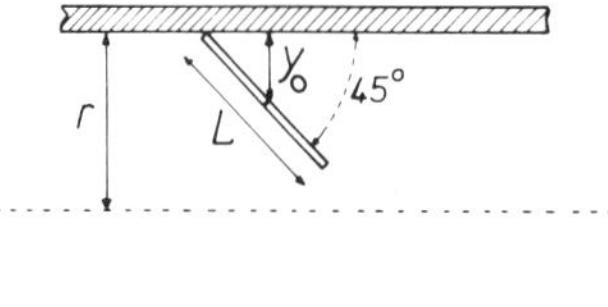

Fig. 33. Model representing rodlike molecule with length L in capillary of radius r, in situation of maximum hydrodynamic shear [Ref. (*15*)]

equation for rodlike molecules having one end in contact with the wall of the capillary and being inclined at 45° to the stream lines (Fig. 33):

$$f = \frac{3\pi\eta}{\log(L/l)} \tag{33}$$

where η is the viscosity of the solvent, and L and l are the length and radius of the rod respectively. For a straight rod in uniform velocity gradient, the distribution of stress is parabolic with the peak at the mitpoint of the molecule. Thus, when the shear stress exceeds the critical value only slightly, the molecule is broken into nearly perfect halves. The stretching force F_{max}, experienced by the molecule is greatest if its orientation is at 45° to the direction of the flow. In a uniform velocity gradient G its value is given by:

$$F_{max} = \frac{f G L^2}{16} \tag{34}$$

In a cylindrical capillary, however, the velocity gradient is not uniform but increases linearly with the distance from the axis. The maximum stretching force will then be given by (*15*):

$$F_{max} = \frac{2 f Q y_0^2}{\pi r^4}\left(r - \frac{2 y_0}{3}\right) \tag{35}$$

where Q is the flow rate (in volume per unit time), r is the radius of the capillary, and y_0 is the distance from the wall of the capillary to that point of the molecule, where the opposite stretching forces are equal; y_0 is given by the root of the equation:

$$3y_0^2 - 6 r y_0 = \frac{L^2}{2} - \frac{3 r L}{\sqrt{2}} \tag{36}$$

This equation shows that the stretching force will be maximal at a point slightly displaced from the center of the molecule.

In their experiments Davison and Levinthal (*15*) degraded DNA with a molecular weight of 1.3×10^8, which corresponds to a contour length of $L = 6.5 \times 10^{-2}$ mm. The radius of the capillary was 0.125 mm, so that the point of breakage will be less than 2×10^{-4} mm from the center of the molecule, which means that the molecule is disintegrated practically at the center of the chain. Using Eqs. (34) and (35) these investigators obtained a value of 11×10^{-4} dynes for the force causing scission of the DNA, which occurs mainly at the C–O bond (*81*). The theoretical tensile strength of the C–O Bond is 8.9×10^{-4} dynes, which is in close agreement with the experimental results.

Harrington and Zimm (*31*), investigated the shear degradation of DNA in several shearing devices, such as a high-pressure capillary, an instrument in which the solution is forced at high pressure through the narrow annulus between a close-fitting piston–cylinder system, sintered glass disks and several high-speed laboratory homogenizers. In all these devices shearing takes place near the walls. In the case of high-speed rotatory homogenizers, the region of high velocity gradient is confined to the boundary layer at the rotating blade, where laminar flow is maintained and in which all degradation occurs. The molecules will be stretched along the streamlines of the flow. A detailed discussion including the evaluation of an equation for the critical force for rupture of macromolecules in such systems is given by Harrington (*32*).

According to Harrington and Zimm (*31*) an *universal* equation to calculate the average force F_{av} sustained by a macromolecule which is submitted to a velocity gradient G is:

$$F_{av} = \frac{G(\eta - \eta_0)}{nz} \tag{37}$$

where η and η_0 are the viscosities of the solution and solvent, n is the number of molecules per unit volume, and z is the mean extension of the molecule in the direction of the velocity gradient. This equation is independent of any model and requires as parameters only the knowledge of the velocity gradient, the viscosity increment of the solution and the extensions of the macromolecule. For a capillary with radius r this equation is transformed into

$$F_{av} = \frac{4M(\eta - \eta_0)}{\pi c N z r^3} \cdot \frac{dV}{dt} \tag{38}$$

where M is the molecular weight of the polymer, c is the concentration, N is Avogadro's number and dV/dt is the volume flow rate of the solution. The values of η and z, which appear in Eqs. (*37*) and (*38*) must be determined at conditions of shear stress of the molecule. In real macromolecules η and z are certainly less in a strong velocity gradient, than the values obtained when the solution is at rest. This point is discussed in detail by Harrington and Zimm (*31*).

Assuming the values of η and z to be the same as those for the solution at rest, these investigators computed the critical average force for DNA having a molecular weight of 1.3×10^8. Using the experimental quantities and Eq. (37), they obtained 2.7×10^{-5} dynes. This is 40 times less than the value of Davison and Levinthal (*15*). The considerable difference between these two results must be due to inadequacies in Davison and Levinthal's molecular model and due to a significant reduction of the parameter z in Harrington and Zimm's equation, because of the extreme shear rates. The theoretical value of 8.9×10^{-4} dynes for the critical strength of a C–O bond represents most probably an *upper* limit, since no solute–solvent interactions have been taken into account. Although the qualitative agreement between theory and experiment may be regarded as satisfactory, the comparison of these data shows clearly the complex nature of the shear degradation process.

6.5. Pulsating Resonant Bubbles

In Section 2.3 it has been shown that during ultrasonic irradiation of liquids, pulsating resonant bubbles can occur. Such vibrating bubbles have a characteristic pattern of eddying liquid surrounding them. It will be shown that even if ultrasonic intensity is too low to cause cavitation in the liquid, this eddying or microstreaming is by itself sufficient to cause disintegration of macromolecules in solution.

It is known through experiments that ultrasonic intensity must be relatively high in order to produce the resonant bubbles, therefore, cavitation cannot be excluded completely. A method, however, has been devised by Hughes and Nyborg (*37*) for causing bubbles of resonant size to vibrate in the liquid at low ultrasonic intensities. To achieve this, the tip of the ultrasonic vibrating horn has been fitted with a series of hemispheric holes having diameter and depth equal to the size of the resonant gas bubbles, which would normally occur in the liquid during irradiation. When the ultrasonic vibrator is immersed into the liquid, air bubbles remain trapped in these holes. Acoustic *microstreaming* around the pulsating bubbles will then occur at low pressure amplitudes of the ultrasonic vibrator, due to the oscillation of the gas bubbles. Using tracer particles, such as Lycopodium spores (diameter 3.5×10^{-3} cm) Pritchard and Peacocke (*79*) investigated the flow field. As a result of the oscillations of the bubble, particles coming from the bulk of the liquid flow radially toward the bubble, change their direction in a small region close to its surface, and then move away from the bubble on a level parallel to the ultrasonic vibrating horn. The situation is shown in Fig. 34. The hemispherical gas bubble resting on

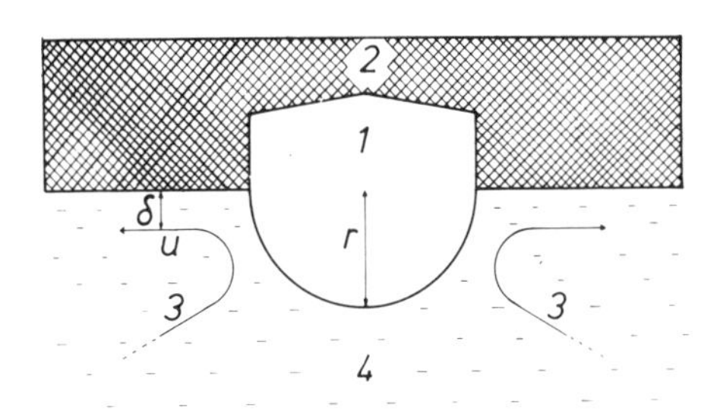

Fig. 34. Pulsating resonant gas bubble of radius r causing acoustic microstreaming near surface of ultrasonic vibrator. *1:* pulsating resonant bubble *2:* ultrasonic vibrator *3:* acoustic microstreaming *4:* bulk of liquid [Ref. (*78*)]

the ultrasonic horn can be regarded as a mechanical resonance system that has one degree of freedom and vibrates almost radially in the ultrasonic field. Assuming a linear velocity gradient generated by microstreaming near the horn boundary, Pritchard and Peacocke arrived at the following expression for its maximum value G_{max}:

$$G_{max} = \frac{u}{\delta} = \frac{A^2 P_x^2}{8\,r^3}\sqrt{\frac{1}{\pi^5 \eta \rho^3 \nu^5}} \tag{39}$$

where u is the maximum flow velocity, δ is the boundary layer thickness, through which u is reduced to zero, A is the pressure amplification factor due to the vibration of the bubble, P_x is the acoustic pressure amplitude at the distance x from the center of the horn, η and ρ are the viscosity and the density of the solution respectively, ν is the ultrasonic frequency and r is the equilibrium resonant radius of the gas bubble. Once the maximum velocity gradient is known, it is possible to calculate the maximum stretching force F_{max}, which a molecule will experience in the flow field. According to their validity any equation from Section 6.3 may be used.

Pritchard *et al.* (*78, 79*) investigated the degradation of calf thymus DNA having a weight-average molecular weight of 6.5×10^6 by the action of pulsating resonant bubbles. For their experimental conditions, *i.e.*, water at atmospheric pressure, the equilibrium-resonant radius, which obeys the equation of Minnaert (*53*), is given by the simplified formula:

$$r = \frac{3}{\nu} \qquad (40)$$

where the radius r is given in millimeters and the frequency ν in kc/sec. At the applied frequency of 20 kc/sec the resonant radius is 0.15 mm. The calculated velocity gradients ranged from 6×10^4 sec^{-1} to 2.5×10^5 sec^{-1}. Furthermore, these investigators computed from the experimental values that the maximum velocity gradient, which was required to produce one break per molecule, was 1.2×10^5 sec^{-1}. Inserting this value into Eq. (33) and (34), which are valid for rigid rodlike molecules, they calculated for DNA ($L = 3\times10^{-4}$ cm, $l = 9\times10^{-8}$ cm) the maximum stretching force F_{max} as 2.3×10^{-3} dynes. This result is in reasonable agreement with the value of 9×10^{-4} dynes to 18×10^{-4} dynes calculated by Davison and Levinthal (*15*) to be necessary to rupture the DNA double helix by the action of hydrodynamic shear forces. Moreover, they found that the macromolecules have the greatest probability of being ruptured at their midpoints, and that degradation proceeded until a limiting molecular weight was reached, which decreased as the ultrasonic intensity was increased. It has been shown by the same investigators that in the absence of vibrating air bubbles, using a smooth ultrasonic horn, DNA was not degraded. In this last case DNA was only degraded if the ultrasonic intensity reached values where cavitation was initiated.

From these theoretical considerations and the experimental investigations, it is evident that the hydrodynamic shear forces generated by microstreaming around resonant bubbles are sufficiently strong to cause degradation of very large macromolecules ($\overline{M}_w > 10^7$) under conditions where transient cavitation is absent.

6.6. Flow Fields Produced by Cavitation

During the collapse of a cavitation bubble strong perturbations are induced in the surrounding liquid and high velocity gradients can be generated. Since a polymer molecule occupies a relatively large volume in solution, the density of a polymer coil is nearly equal to the density of the pure solvent, so that the former will normally follow the flow of the solvent. If the solution is moderately concentrated, entanglements between different polymer molecules may occur. As the polymer molecules will generally not move with the same velocity, because of the inhomogeneous flow fields, shear stresses are created within the molecule due to the entanglements of different chain segments. If these stresses are greater than the critical tensile strength of the chemical bonds involved, the molecules are disrupted. Both monomolecular and bimolecular reactions may then occur. These cases have been discussed in detail by Okuyama and Sata (*68*), but due to the complexity of the phenomena, no equations for shear stresses have yet been derived. The following considerations are therefore restricted to very dilute polymer solutions, in which no association of different polymer molecules is supposed to occur.

A simple degradation model is proposed by Thomas (*103*). During the collapse of a cavity the polymer coil moves with the liquid toward the center of the cavity. Since the flow of the solvent is *inhomogeneous,* the side of the polymer coil near the collapsing cavity will move with a higher velocity than the side turned away from the cavity. In this way stresses are generated within the polymer coil, which will distort it from its initial shape and unfold it until a geometry is reached, which is incapable of further extension. At this time it is assumed that the polymer molecule has a rodlike shape with length L and diameter l. The extended polymer will flow through the solution at a velocity that lies between the velocity of the solvent and that of the two ends. The situation is depicted in Fig. 35. At the point of maxi-

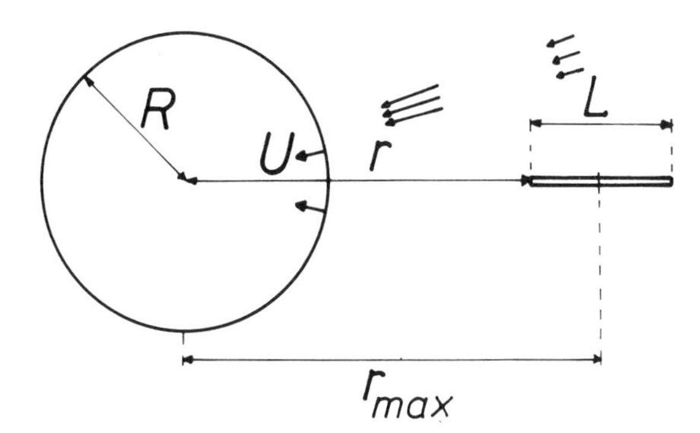

Fig. 35. Degradation model assuming inhomogenous flow fields [Ref. (*103*)]

mum stress, at distance r_{max} from the center of the cavity, the velocity of the polymer is equal to the velocity of the solvent. Assuming that the molecule resembles a string of spherical beads, the friction force on a segment may be calculated by Stoke's law. Integrating this friction force from r to r_{max}, Thomas obtained:

$$F = \frac{3\pi \, \eta l \, U R^2 \, (r_{max} - r)^2}{a \, r \, r_{max}^2} \tag{41}$$

where η is the viscosity of the solvent, U is the velocity of collapse of the cavity, R is the radius of the cavity, a is the length of a monomer unit and r is the shortest distance form the center of the cavity to the polymer chain. Integrating the friction force from r_{max} to $r + L$ and equating the two forces, one obtains:

$$r_{max} = \sqrt{r\,(r + L)} \tag{42}$$

For $r \gg L$, which is reasonable for most degradation events, it is readily seen that the position of r_{max} is *near the center* of the polymer molecule. The stretching force reduces then to:

$$F = \frac{3\pi \, \eta l \, U R^2 L^2}{4 \, a \, r^3} \tag{43}$$

This equation contains three variables, U, R and r for which it is impossible to give numerical values, since all three variables change greatly during the collapse of the cavitation bubble.

Suppanz (*100*) used this model and by including energy balances and applying the equation of continuity of fluid flow, he deduced an equation for the friction force acting on the polymer molecule, in terms of the physical constants characterizing the polymer and the solvent, as a function of the initial radius of the cavitation bubble. An explicit solution of the equation of motion obtained is not possible, but numerical values for different conditions could be computed. Assuming fully extended dextran polymers in solution, Suppanz carried out such computations for molecular weights from 50000 to 150000, which correspond to a contour length of approximately 2×10^{-5} cm to 6×10^{-5} cm, in water for different initial radii R_0 of the cavitation bubbles. The polymer molecule is considered to be in the vicinity of the cavitation bubble at the initial stage of collapse. Typical results are shown in Fig. 36 from which it follows that degradation is possible, assuming an initial radius of the cavitation bubble of 1.5×10^{-3} cm and the polymer has a molecular weight of 100000 or greater.

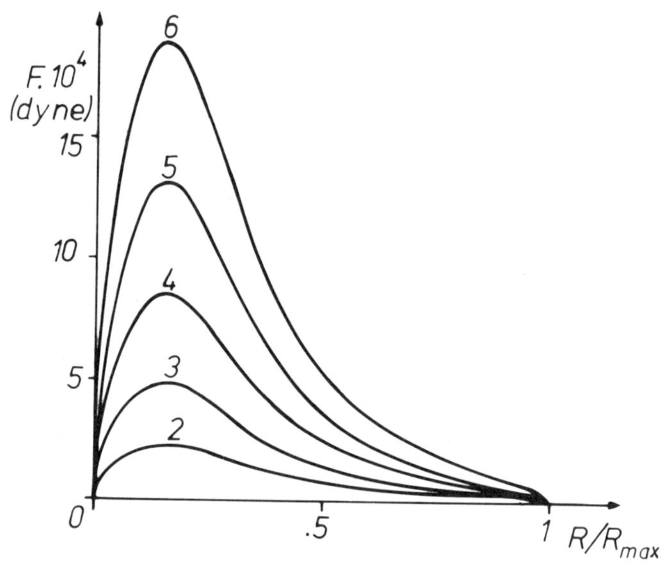

Fig. 36. Calculated stretching force (F) acting on polymer molecule as function of radius (R/R_{max}) of cavitation bubble. R_{max}: maximum radius of cavity, assumed in this case to be 1.5×10^{-3} cm. Parameter of curves is contour length x 10^5 (in cm) of the polymer [Ref. (*100*)]

Thomas (*103*) showed that a *linear* dependence of the degradation rate constants on the degree of polymerization can be deduced from this model. This has been repeatedly confirmed by experimental results (*2, 97*). A detailed discussion of the model modified by Suppanz (*100*), however, reveals that a dependence of the degradation constants on the molecular weight raised to a power of 1.5 can also be explained by the mechanism. From experimental results of Glynn *et al.* (*22, 23*) a value of 1.25 for this exponent was best for degradation of polystyrene in tetrahydrofuran.

Okuyama and Hirose (*70*) proposed a slightly different mechanism, considering that the stretching force acts only on a partial chain segment. At the contraction step of a cavitation bubble, a partial segment of the chain takes a linearly extended configuration due to friction forces between the solvent molecules and the polymer chain, caused by the flow. This partial chain is subjected to a tensile force, which is maximum at the bond in the middle of this partial chain (Fig. 37). From these ob-

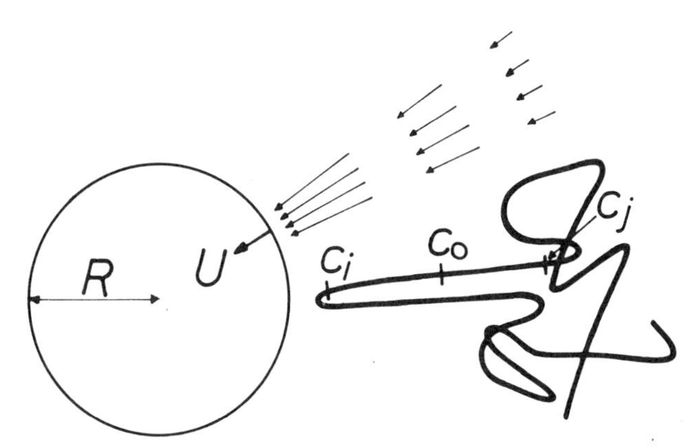

Fig. 37. Degradation model assuming inhomogenous flow fields and stretched partial polymer chain. C_i–C_j: partial chain, C_0: central atom of the partial chain [Ref. (*70*)]

servations an expression for the final value of the degree of polymerization g, after an infinite time of irradiation was deduced:

$$g = \sqrt{\frac{(F_{C-C})_{max}}{r_0^2 \cdot \zeta^* (U/R)_{max}}} \tag{44}$$

where $(F_{C-C})_{max}$ is the critical tensile strength of the C–C bond considered r_0 is the equilibrium distance between the atoms forming the bond, ζ^* is the microstreaming friction coefficient between the polymer molecule and the solvent, per unit length of the polymer, and $(U/R)_{max}$ is the maximal value of the relation of the surface velocity of the contracting bubble ($-U$) and its radius R. By comparing the experimental values of g with those calculated for different cavitation conditions, Okuyama and Hirose tried to decide whether gaseous cavitation, which consists of vibrating bubbles having a radius of about 10^{-2} cm filled with gas or transient cavitation bubbles with an radius of about 10^{-5} cm, were the cause of degradation. At an ultrasonic frequency of 500 kc/sec and an acoustic pressure amplitude of 4 bars these investigators obtained a limiting degree of polymerization of 250 for polystyrene in toluene. The microscopic friction coefficient was determined as 0.01 dyne sec/cm^2. For a gas-filled cavity $(U/R)_{max}$ is of the order of $1-3 \times 10^6$ sec^{-1}, which according to Eq. (44) gives a value of 8×10^3 to 12×10^3 for the limiting degree of polymerization, which is many times greater than the experimental value. For a transient cavity $(U/R)_{max}$ is of the order of 1.8×10^{10} sec^{-1}, which gives a limiting degree of polymerization of 100. From these results, these investigators concluded that the limiting degree of polymerization and the degradation kinetics of the reaction is determined by the collapse of *transient cavities*.

6.7. Shock Waves Produced by Cavitation

The intense shock wave radiated from a cavitation bubble at the final stage of the collapse, is undoubtedly the cause of the most severe reactions that occur in connection with high-intensity ultrasonics in liquids. It has been shown in Section 2.4 that a shock wave can be considered in this case as a rapid pressure rise followed by an extremely sharp exponential pressure drop. Now it will be shown that this shock wave too, is capable of causing the scission of macromolecules that lie in its path.

Gooberman (*26*) proposed a mechanism that explains the rupture of polymer molecules as a result of this action. During the period of pressure rise, the solvent is compressed, and a larger number of solvent molecules will then be present in the volume unit. On assumption that the polymer molecule, which in solution has the shape of a more or less expanded coil, does not change its configuration appreciably during the time of compression, solvent molecules will flow inside the volume enclosed by the polymer coil. During the subsequent rapid pressure drop, the liquid expands and the solvent molecules flowing then out of the polymer coil will generate stresses within the macromolecule due to friction forces. During this period of time, the polymer coil is assumed to retain again its configuration. This assumption is plausible, since configurational relaxation times of macromolecules are of the order of 10^{-3} sec, which is many orders of magnitude greater than the time required for a shock wave to travel across the polymer coil, which is usually of the order of 10^{-10} sec. This is supported by the experimental evidence that the overall degradation rate is a function of the equilibrium configuration of the macromolecule (*1, 2*).

For simplicity the shock wave is assumed to be spherical for all qualitative interpretations. In this case the solvent flow will be in the direction of an extended radius of the shock wave. If the solution is sufficiently dilute that a given macromolecule can be considered completely isolated from all others, it will move with the solvent so that the net force acting on it due to the solvent flow will be zero. This means that the velocity of the center of mass of the molecule is equal to the velocity of the adjacent solvent. During the expansion period of the solvent, the path moved by a solvent molecule relative to the center of mass of the polymer and in a direction parallel to an extended radius of the shock wave, in a certain interval of time, is proportional to its distance from the center of mass. This means that the solvent velocity relative to the polymer chain will increase with the distance from the center of mass of the polymer coil, generating in this way a velocity gradient across the polymer chain.

If the stress created within the macromolecule as a result of the friction forces between the solvent and the segments of the polymer chain due to this velocity gradient is sufficiently large, rupture of chemical bonds can occur. It is therefore necessary to study the stress distribution along the polymer chain, which will depend on the orientation of the segments relative to the shock wave. Following the argumentation of Gooberman (*26*), the maximum stress will be situated in a plane normal to the direction of propagation of the shock wave, and which passes through the center of mass of the polymer chain. The section of the polymer molecule between this plane and the chain end is assumed to have the configuration of a random coil. According to this model the bond most likely to break will be *near* the center of mass of the molecule. Because of the variable configuration of polymers in solution and the many possibilities of orientation of the molecule relative to the solvent flow, the bond most likely to break is not necessarily that at the center of the chain. In the case of fairly rigid linear macromolecules or in molecules extended in a flow field, rupture should, however, occur at its geometric center.

Gooberman (*26*) performed such calculations for polystyrene in benzene. Assuming that the macromolecule has the form of a random coil containing $2n + 1$ segments of length b he obtained an equation for the force F acting on the center bond:

$$F = 0.67\, \zeta\, G\, b\, n^{1.5} \qquad (45)$$

where G is the velocity gradient and ζ the friction coefficient of the segment in the solvent. The velocity gradient G, generated during the expansion of the compressed solvent, depends on its equation of state, which relates pressure and volume. For benzene as solvent Gooberman arrived at the following equation:

$$G = \frac{0.18\, \zeta\, P_m}{\gamma\, \theta\, (P_m + 1000)^{1.09}} \qquad (46)$$

where P_m is the peak pressure of the shock wave near the polymer molecule, θ is the time constant of pressure decay, and γ is the time spread parameter, which takes into account the reduction in the rate of pressure drop as the shock wave advances. Eqs. (45) and (46) contain four quantities, ζ, θ, γ, and P_m for which accurate values are unknown. The value of ζ can be estimated from Stoke's law, but θ and γ depend on the minimum radius achieved by the collapsing cavitation bubble, which is again an unknown quantity. Taking 5×10^{-5} cm for the minimum radius at an ultrasonic frequency of 200 kc/sec and an acoustic pressure amplitude of 2 bars, Gooberman calculated that a value for P_m of 650 bars would be required to break a polystyrene molecule with a molecular weight of 10^5 ($n \approx 500$). The critical strength of a C–C bond has been assumed to be 5.54×10^{-4} dynes. This pressure corresponds to a pressure amplitude of the shock wave of 1640 bars at the point of collapse of the cavitation bubble. Because of the many simplifications and arbitrary values in Gooberman's computations, this value can be regarded as a rough approximation only. Nevertheless, it falls well within the limits of the pressure amplitudes of shock waves generated by ultrasonic cavitation. The degradation of polymers in solution by ultrasound is attributed also to the action of shock waves by other investigators (*93, 95*).

From Gooberman's observations, it follows that below a molecular weight of 10^6 the rate constants of degradation of polystyrene in benzene are proportional to the molecular weight raised to a power of 2.82 (*26*). Beyond this molecular weight the rate constants increase at a slower rate. These results, however, have not been confirmed by other investigators. Relating the peak pressure generated during the collapse of a cavitation bubble to the ultrasonic pressure amplitude and to the value of n in Eq. (45), Gooberman stated that the limiting molecular weight of polystyrene in benzene should be inversely proportional to the acoustic pressure amplitude raised to a power of 2.33. This means that the limiting molecular weight is roughly inversely proportional to the ultrasonic intensity, which has been experimentally confirmed in some cases.

Gooberman and Lamb (*27, 28*) also determined experimentally the molecular-weight distributions of the degraded products using a turbidimetric technique. Their results indicate that the initial material, which had an average molecular weight of about 2×10^5, tends to split near the *midpoint* of the chain. At low ultrasonic intensities (3.4 watts/cm^2, 200 kc/sec), the degradation products did not contain fragments with a molecular weight smaller than 50000. Some uncertainties, however, cannot be neglected in these results, too.

7. Concluding Remarks

It is now generally agreed that ultrasonic degradation of polymers in solution is of mechanical nature, and that the stresses set up in the polymer molecule are caused by the friction forces generated by the relative movement of the molecules of solvent and the polymer molecule, as a result of the collapse of cavitation bubbles. Many experimental investigations demonstrate clearly that ultrasonic degradation is not thermal in origin. If thermal degradation is considered, large differences in the rate constants should occur if the solution is saturated with a monoatomic gas, which gives the highest temperatures, or with polyatomic gases, which give the lowest temperatures. In practice no differences in degradation rates are found, which proves that degradation does *not* take place *inside* the cavitation bubbles. Moreover, it has been found that the rate of ultrasonic degradation decreases with increasing temperature. This *negative* temperature coefficient, which has been found for a large number of polymers and using a variety of shear degradation methods, is a definite criterion of a mechanochemical reaction. The prominent *inefficiency* of bond rupture by ultrasound is also a characteristic feature of mechanochemical reactions and is generally valid for polymer systems where mechanical degradation methods are considered. During irradiation of polymer solutions with ultrasound only a factor of about 10^{-5} of the total energy input is used to break chemical bonds.

The dependence of the rate constants of degradation on the molecular weight of the polymer is also a consequence of the mechanical nature of the ultrasonic degradation process. Larger polymer molecules present more resistance to flow, and therefore accumulate greater shear forces, thus leading more frequently to rupture than shorter polymer chains. The existence of a *limiting* degree of polymerization is another necessary consequence of this, since there is always a minimum chain length below which the forces set up within the molecule cannot exceed the bond strength, and consequently the chain cannot break. If degradation were thermal or chemical, the dependence of the degradation rate constant on the molecular weight of the polymer molecule could hardly be understood; the existence of a limiting chain length could not be explained at all. Schmid *et al.* (*90*) pointed out very early the fundamental differences between ultrasonic and chemical or thermal degradation of polymers. The fact that an initially broad molecular-weight distribution becomes more homogeneous during insonation may be considered further proof for the mechanical nature of ultrasonic degradation, since this effect is not observed during thermal or chemical degradation of polymers.

The experimental result that in dilute solutions the degradation reaction follows a first-order kinetic indicates that the polymer molecules are degraded independently from each other. The result of Fig. 26, that the slopes of the lines which give the dependence of the rate constants of degradation on the molecular weight of the polymer, are proportional to the rate constants themselves, demonstrates that each monomer unit contributes, within a definite range of molecular weights, with the same portion to the degradation constant of the whole molecule. This proves that the molecule is submitted to tensile stress *before* rupture takes place. The same conclusion may be drawn from the finding that the polymer molecule does not break statistically, but preferentially at regions close to its midpoint. The fact that the line in Fig. 26 is independent of the intensity of ultrasound, of the intensity of cavitation, of temperature, of the solvent and other experimental conditions, proves that the mechanism of degradation is *the same* in all the cases considered in this figure. Since the chemical properties of the solvents used are of very different nature, and degradation, however, follows the same laws, more evidence for a *mechanical degradation mechanism* is given.

The degradation mechanisms described in Chapter 6 demonstrate that the mechanical stress on the polymer molecule is the primary cause of the bond breakage. Polymer molecules are distorted and stretched as they enter the area of high velocity gradients generated by collapsing cavitation bubbles. At the final stage of collapse, the shock wave radiated from the cavity generates the stresses within the polymer molecule and constitutes most probably the ultimate cause of rupture. Valuable information could be supplied, if the degree of extension of the polymer coil under stress were known, and a comparison of this value to the normal contour length of the chain could be made.

Many investigators who studied the effect of solvents on the reaction rate of shear degradation processes, found that at moderate concentrations the polymer molecules are degraded more easily in poor solvents. The experiments of Johnson and Price (*46*) and Nakano *et al.* (*61–63*) indicated that maximum degradation was obtained in poor solvents, where the polymer coils are contracted and at low temperatures. This is contrary to the effect of solvents on ultrasonic degradation, where it has been shown (*1, 2*) that degradation is more pronounced in good solvents. Since at shear degradation the friction forces act on the surface of the polymer coil and depend largely on entanglements (*10*), whereas considering shock waves generated by ultrasonic cavitation, the stresses are set up *within* the polymer coil, these results are understandable and favor a mechanism of shock waves, similar to that proposed by Gooberman (*26*) for explaining ultrasonic degradation. This also explains the fact that suspensions of very fine particles of polymers in nonsolvents are not degraded by ultrasound.

The physicochemical properties of the solvent, which have an important effect on the breaking force, are not yet properly understood. Polymer–solvent interactions are always present, and definitely affect the tensile strength of the bonds involved, whereas the critical stresses calculated in Section 6.2, although entirely valid for gas-phase processes, are not applicable in solution without many restrictions. This is indicated by the fact that different polymers with the same chemical bonds along the main chain have different rate constants of degradation in the same solvent. The hydrodynamic and mechanical effects in the chain–scission process are certainly far more involved than the simple stretching of chemical bonds to their point of rupture.

8. References

1. Alexander, P., Fox, M.: The role of free radicals in the degradation of high polymers by ultrasonic and by high-speed stirring. J. Polym. Sci. **12**, 533 (1954).
2. Basedow, A. M., Ebert, K. H.: Zum Mechanismus des Abbaus von Polymeren in Lösung durch Ultraschall. Makromol. Chem. **176**, 745 (1975).
3. Basedow, A. M.: Untersuchungen über den Abbau von Dextran durch Ultraschall in verschiedenen Lösungsmitteln. Thesis, University of Heidelberg, Faculty of Chemistry 1973.
4. Basedow, A. M., Ebert, K. H., Ederer, H., Hunger, H.: Die Bestimmung der Molekulargewichtsverteilung von Polymeren durch Permeationschromatographie an porösem Glas. Makromol. Chem. **177**, 1501 (1976).
5. Bergmann, L.: Der Ultraschall. Hirzel Verlag (1954).
6. De Boer, J. H.: The influence of van der Waals forces and primary bonds on binding energy, strength and orientation. Trans. Far. Soc. **32**, 10 (1936).
7. Bohn, L.: Schalldruckverlauf und Spektrum bei der Schwingungskavitation. Acustica **7**, 201 (1957).
8. Blandamer, M. J.: Introduction to chemical ultrasonics. Academic Press 1973.
9. Bradbury, J. H., O'Shea, J. M.: The effect of ultrasonic irradiation on proteins. Austral. J. Biol. Sci. **26**, 583 (1973).
10. Bueche, F.: Mechanical degradation of high polymers. J. Appl. Polymer Sci. **4** (10), 101 (1960).
11. Butyagin, P. Y.: Kinetics and nature of mechanochemical reactions. Russ. Chem. Rev. **40** (11), 901 (1971).
12. Chandra, S., Roy-Chowdhury, P., Biswas, A. B.: Ultrasonic degradation of sol rubber in solution. J. Appl. Polym. Sci. 8, 2653 (1964).
13. Chandra, S., Roy-Chowdhury, P., Biswas, A. B.: Ultrasonic degradation of macromolecules in solution. J. Appl. Polym. Sci. **10**, 1089 (1966).
14. Chendke, P. K., Fogler, H. S.: Second-order sonochemical phenomena: extensions of previous work and applications in industrial processing. Chem. Eng. J. **8**, 165 (1974).
15. Davison, P. F., Levinthal, C.: Degradation of deoxyribonucleic acid under hydrodynamic shearing forces. J. Mol. Biol. **3**, 674 (1961).
16. Davison, P. F., Freifelder, D.: Studies in the sonic degradation of deoxyribonucleic acid, Biophys. J. **2**, 235 (1962).
17. Él'Piner: Ultrasound: Physical, Chemical and Biological Effects. Consultants Bureau (1964).
18. Él'Tsefon, B. S., Berlin, A. A.: Investigations in the mechanochemistry of polymers XIII. Vysokomolekul. Soedin. **4**, 1033 (1962).
19. Flügge, S.: Encyclopaedia of Physics 11/2 – 2. Springer Verlag (1962).
20. Flynn, H. G.: Physics of Acoustic Cavitation in Liquids. In: Physical acoustics Vol. 1-B. Mason, W. (ed.), p. 51. Academic Press 1964.
21. Fujiwara, H., Okazaki, K., Goto, K.: Mechanochemical reaction of polymers by ultrasonic irradiation I. J. Polym. Sci. **13**, 953 (1975).
22. Glynn, P. A., Van der Hoff, B. M., Reilly, P. M.: A general model for prediction of molecular weight distributions of degraded polymers. J. Macromol. Sci. **A6** (8) 1653 (1972).
23. Glynn, P. A., Van der Hoff, B. M.: Degradation of polystyrene in solution by ultrasonation. A molecular weight distribution study. J. Macromol. Sci. **A7** (8), 1695 (1973).
24. Glynn, P. A., Van der Hoff, B. M.: The rate of degradation by ultrasonation of polystyrene in solution. J. Macromol. Sci. **A8** (2), 429 (1974).
25. Gooberman, G.: Ultrasonics: Theory and application. Hart Publishing Co. 1969.
26. Gooberman, G.: Ultrasonic degradation of polystyrene. Part 1. J. Polym. Sci. **42**, 25 (1960).
27. Gooberman, G., Lamb, J.: Ultrasonic degradation of polystyrene. Part 2. J. Polym. Sci. **42**, 35 (1960).
28. Gooberman, G.: Ultrasonic degradation of polystyrene. Part 3. J. Polym. Sci. **47**, 229 (1960).
29. Gueth, W.: Zur Entstehung der Stoßwellen bei der Kavitation. Acustica **6**, 526 (1956).
30. Gueth, W., Mundry, E.: Kinematographische Untersuchungen der Schwingungskavitation. Acustica **7**, 241 (1957).

31. Harrington, R. E., Zimm, B. H.: Degradation of polymers by controlled hydrodynamic shear. J. Phys. Chem. **69** (1), 161 (1965).
32. Harrington, R. E.: Degradation of polymers in high speed rotary homogenizers. J. Polym. Sci. **4**, 489 (1966).
33. Henglein, A.: Die Auslösung und der Verlauf der Polymerisation des Acrylamids unter dem Einfluß von Ultraschallwellen. Makromol. Chem. **14**, 15 (1954).
34. Henglein, A.: Die Reaktion des DPPH mit langkettigen freien Radikalen. Makromol. Chem. **15**, 188 (1955).
35. Henglein, A.: Die Kombination von freien makromolekularen Radikalen, die durch Ultraschallabbau von Polymethacrylsäuremethylester und von Polystyrol gebildet werden. Makromol. Chem. **18**, 37 (1956).
36. Heymach, G. J., Jost, D. E.: The alteration of molecular weight distributions of polymers by ultrasonic energy. J. Polym. Sci. **C25**, 145 (1968).
37. Hughes, D. E., Nyborg, W. L.: Cell disruption by ultrasound. Science **138**, 108 (1962).
38. Jellinek, H. H., White, G.: The degradation of long-chain molecules by ultrasonic waves. 1. J. Polym. Sci. **4** (6), 745 (1951).
39. Jellinek, H. H., White, G.: The degradation of long-chain molecules by ultrasonic waves. 2. J. Polym. Sci. **6** (6), 757 (1951).
40. Jellinek, H. H., White, G.: The degradation of long-chain molecules by ultrasonic waves. 3. J. Polym. Sci. 7 (1), 21 (1951).
41. Jellinek, H. H., White, G.: The degradation of long-chain molecules by ultrasonic waves. 4. J. Polym. Sci. 7 (1), 33 (1951).
42. Jellinek, H. H., Brett, H. W.: Degradation of long-chain molecules by ultrasonic waves. 5. J. Polym. Sci. **13**, 111 (1954).
43. Jellinek, H. H., Brett, H. W.: Degradation of long-chain molecules by ultrasonic waves. 6. J. Polym. Sci. **21**, 535 (1956).
44. Jellinek, H. H.: Degradation of long-chain molecules by ultrasonic waves. 7. J. Polym. Sci. **22**, 149 (1956).
45. Jellinek, H. H.: Degradation of long-chain molecules by ultrasonic waves. 8. J. Polym. Sci. **37**, 485 (1959).
46. Johnson, W. R., Price, C. C.: Shear degradation of vinyl polymers in dilute solution by high-speed stirring. J. Polym. Sci. **45**, 217 (1960).
47. Lippincott, E. R., Schroeder, R.: General relation between potential energy and internuclear distance for diatomic and polyatomic molecules. J. Chem. Phys. **23** (6), 1131 (1955).
48. Marique, L. A., Houghton, G.: Analog computer solution of the modified Rayleigh equation and parameters affecting cavitation. Can. J. Chem. Eng. **122** (1962).
49. Mark, H.: Some applications of ultrasonics in high-polymer research. J. Acoust. Soc. Am. **16** (3), 183 (1945).
50. Mason, W.: Physical Acoustics Vol. II-A, B. Academic Press 1965.
51. Mellen, R. H.: An experimental study of the collapse of a spherical cavity in water. J. Acoust. Soc. Am. **28** (3), 447 (1956).
52. Melville, H. W., Murray, A. J.: The ultrasonic degradation of polymers. Trans. Far. Soc. **46**, 996 (1950).
53. Minnaert, M.: On musical air-bubbles and the sound of running water. Phil. Magaz. **16** (7), 235 (1933).
54. Mostafa, M. A.: Degradation of addition polymers by ultrasonic waves. 1. J. Polym. Sci. **22**, 535 (1956).
55. Mostafa, M. A.: The degradation of addition polymers by ultrasonic waves. 2. J. Polym. Sci. **27**, 473 (1958).
56. Mostafa, M. A.: Degradation of addition polymers by ultrasonic waves. 3. J. Polym. Sci. **28**, 499 (1958).
57. Mostafa, M. A.: Degradation of addition polymers by ultrasonic waves. 4. J. Polym. Sci. **28**, 519 (1958).
58. Mostafa, M. A.: Degradation of addition polymers by ultrasonic waves. 5. J. Polym. Sci. **33**, 295 (1958).

59. Mostafa, M. A.: Degradation of addition polymers by ultrasonic waves. 6. J. Polym. Sci. **33**, 311 (1958).
60. Mostafa, M. A.: A Mechanism of degradation of long-chain molecules by ultrasonic waves. J. Polym. Sci. **33**, 323 (1958).
61. Nakano, A., Minoura, Y., Kasuya, T., Kawamura, S.: Degradation of poly(ethylene oxide) by high-speed stirring. J. Polym. Sci. **5**, 125 (1967).
62. Nakano, A., Minoura, Y.: Degradation of polymers by high-speed stirring. J. Appl. Polym. Sci. **15**, 927 (1971).
63. Nakano, A., Minoura, Y.: Effect of solvents on the degradation of polymers by high-speed stirring. J. Appl. Polym. Sci. **16**, 627 (1972).
64. Noltingk, B. E.: The effects of Intense Ultrasonics in Liquids. In: Encyclopaedia of physics. Flügge, S. (ed.), p. 259. Springer-Verlag 1962.
65. Noltingk, B. E., Neppiras, E. A.: Cavitation produced by ultrasonics. Proc. Phys. Soc. **B63**, 674 (1950).
66. Noltingk, B. E., Neppiras, E. A.: Cavitation produced by ultrasonics: theoretical conditions for the onset of cavitation. Proc. Phys. Soc. **B64** (12), 1032 (1951).
67. Nosov, V.: Ultrasonics in the chemical Industry. Consultants Bureau (1965).
68. Okuyama, M., Sata, N.: Der Ultraschallabbau langkettiger Moleküle. 1. Z. Elektrochem. **58** (3), 197 (1954).
69. Okuyama, M.: Der Ultraschallabbau langkettiger Moleküle und der Mechanismus der Kavitation. 2. Z. Elektrochem. **59** (6), 565 (1955).
70. Okuyama, M., Hirose, T.: Mechanics of ultrasonic degradation of linear high polymer and ultrasonic cavitation. J. Appl. Polym. Sci. **7**, 591 (1963).
71. Okuyama, M., Hirose, T.: Physico-chemical approach to ultrasonic cavitation. Kolloid Zeitsch. **226** (1), 70 (1967).
72. Ovenall, D. W., Hastings, G. W.: The degradation of polymer molecules in solution under the influence of ultrasonic waves. 1. J. Polym. Sci. **33**, 207 (1958).
73. Ovenall, D. W., Allen, P. E., Burnett, G. M., Hastings, G. W., Melville, H. W.: The degradation of polymer molecules in solution under the influence of ultrasonic waves. 2. J. Polym. Sci. **33**, 213 (1958).
74. Ovenall, D. W.: Ultrasonic degradation of polymer molecules in solution: some comments on recent papers. J. Polym. Sci. **42**, 455 (1960).
75. Porter, R. S., Cantow, M. J. Johnson, J. F.: Sonic degradation of polyisobutylene in solution. J. Appl. Phys. **35** (1), 15 (1964).
76. Porter, R. S., Cantow, M. J. Johnson, J. F.: Polymer degradation. 5. Changes in molecular weight distribution during sonic irradiation of polyisobutene. J. Appl. Polym. Sci. **11**, 335 (1967).
77. Porter, R. S., Casale, A.: The mechanochemistry of high polymers. Rubber Chem. Techn. **1971**, 534.
78. Pritchard, N. J., Hughes, D. E., Peacocke, A. R.: The ultrasonic degradation of biological macromolecules under condition of stable cavitation. 1. Biopolymers **4**, 259 (1966).
79. Pritchard, N. J., Peacocke, A. R.: The ultrasonic degradation of biological macromolecules under condition of stable cavitation. 2. Biopolymers **6**, 605 (1968).
80. Rayleigh, L.: On the pressure developed in a liquid during the collapse of a spherical cavity. Phil. Magaz. **34**, 94 (1917).
81. Richards, O. C., Boyer, P. D.: Chemical mechanism of sonic, acid, alkaline and enzymic degradation of DNA. J. Mol. Biol. **2**, 327 (1965).
82. Schmid, J.: Kinematographische Untersuchungen der Einzelblasen-Kavitation. Acustica **9**, 321 (1959).
83. Schmid, G., Rommel, O.: Zerreißen von Makromolekülen mit Ultraschall. Z. Physikal. Chem. **185** (2), 97 (1939).
84. Schmid, G.: Zur Kinetik der Ultraschalldepolymerisation. Z. Physikal. Chem. **186** (3), 113 (1940).
85. Schmid, G., Rommel, O.: Zerreißen von Makromolekülen mit Ultraschall. Z. Elektrochem. **45** (9), 659 (1939).

86. Schmid, G.: Zerreiben von Molekülen. Versuch einer Erklärung der depolymerisierenden Wirkung von Ultraschallwellen. Phys. Z. **41**, 326 (1940).
87. Schmid, G., Beuttenmüller, E.: Ultraschallbeitrag zur Frage der Biegsamkeit der Makromoleküle. Z. Elektrochem. **49** (4–5), 325 (1943).
88. Schmid, G., Beuttenmüller, E.: Der Einfluß der Temperatur auf den Abbau von linearen Makromolekülen mit Ultraschall. Z. Elektrochem. **50** (9–10), 209 (1944).
89. Schmid, G., Poppe, W.: Die Frequenzunabhängigkeit des Ultraschallabbaus von Makromolekülen. Z. Elektrochem. **53** (1), 28 (1949).
90. Schmid, G., Paret, G., Pfleiderer, H.: Die mechanische Natur des Abbaus von Makromolekülen mit Ultraschall. Kolloid-Z. **124** (3), 150 (1951).
91. Schmid, G., Schneider, C., Henglein, A.: Die Veränderung der Polymer-Einheitlichkeit beim Ultraschall-Abbau von Polymethacrylsäuremethylester. Kolloid-Z. **148** (1–2), 73 (1956).
92. Schoon, T. G., Kretschmer, R.: Versuch einer Deutung bei der elektronenmikroskopischen Untersuchung von festen Polymeren. Kolloid-Z. **211** (1–2), 53 (1966).
93. Schoon, T. G., Rieber, G.: Theorie des Ultraschall-Abbaus von Polymeren in Lösung auf der Grundlage des Perlschnurmolekül-Modells. Angew. Makromol. Chem. **15** (226), 263 (1971).
94. Schoon, T. G., Rieber, G.: Ultraschallabbau und Mikromorphologie amorpher Hochpolymeren. Angew. Makromol. Chem. **23** (307), 43 (1972).
95. Schoon, T. G., Rieber, G.: Theorie des Ultraschallabbaus auf der Grundlage des Perlschnurmolekül-Modells. Angew. Makromol. Chem. **49** (704), 23 (1976).
96. Shaw, M. T., Rodriguez, F.: Ultrasonic degradation of polysiloxane solutions. J. Appl. Polym. Sci. **11**, 991 (1967).
97. Smith, W. B., Temple, H. W.: Polymer studies by gel permeation chromatography. 4. The degradation of polystyrene by ultrasonics and by benzoyl peroxide. J. Phys. Chem. **72** (13), 4613 (1968).
98. Stuart, H.: Molekülstruktur. Springer-Verlag 1967.
99. Stuart, H.: Die Physik der Hochpolymeren. Vol. 1. Springer-Verlag 1952.
100. Suppanz, N.: Über Kinetik und Mechanismus des Abbaus von Polymeren in Lösung mit Ultraschall, Thesis, University of Heidelberg, Faculty of Chemistry 1972.
101. Thomas, B. B., Alexander, W. J.: Ultrasonic degradation of cellulose nitrate. 1. J. Polym. Sci. **15**, 361 (1955).
102. Thomas, B. B., Alexander, W. J.: Ultrasonic degradation of cellulose nitrate. 2. J. Polym. Sci. **25**, 285 (1957).
103. Thomas, J. R.: Sonic degradation of high polymers in solution. J. Phys. Chem. **63**, 1725 (1959).
104. Thomas, J. R., De Vries, L.: Sonically induced heterolytic cleavage of polymethylsiloxane J. Phys. Chem. **63**, 254 (1959).
105. Weissler, A.: Depolymerization by ultrasonic irradiation: the role of cavitation. J. Appl. Phys. **21**, 171 (1950).
106. Weissler, A.: Cavitation in ultrasonic depolymerization. J. Appl. Phys. **23**, 370 (1951).
107. Wilke, G., Altenburg, K.: Ultraschallabbau von Hochpolymeren. Plaste und Kautschuk **3** (10), 219 (1956).
108. Wilke, G., Altenburg, K.: Ultraschallabbau von Hochpolymeren. Plaste und Kautschuk **3** (11), 257 (1956).

Received June 14, 1976

Author Index Volumes 1-22

J. J. Bikerman

Foams

79 figures. IX, 337 pages. 1973
(Applied Physics and Engineering, Vol. 10)

There has long been a need for a source of thorough and critically evaluated information on liquid foams. This monograph answers that need. It will be of value to physical chemists and chemical engineers in academic life as well as research and plant scientists working in numerous industries (such as the manufacture of paper, beer, cosmetics, and antibiotics) and to those employed by many public utilities. It will prove an excellent supplementary text for graduate courses in the physical chemistry of surfaces. The following sources of foam are dealt with in detail: natural water, boiling water ("carry-over" in boilers), pickling liquor in ferrous metallurgy, metallurgical slags, petroleum products, paper intermediates, sugar intermediates, beer, detergents, and fermentation broth for antibiotics. The use of foam for substance separation (or foam fractionation) has recently become an efficient branch of chemical engineering and may be applied to sewage treatment and the disposal of radioactive waste. Other uses include retarding evaporation in reservoirs, accelerating heat and mass transfer, petroleum recovery, and heterogeneous chemical reactions.

Electrons in Fluids

The Nature of Metal-Ammonia Solutions
Editors: J. Jortner, N. R. Kestner
With contributions by numerous experts
271 figures, 59 tables. XII, 493 pages. 1973

This full and up-to-date account of the chemical and physical properties of electrons in polar, nonpolar, and dense fluids includes contributions from both theoretical and experimental chemists and physicists, thus clearly indicating the interdisciplinary nature of this field.

G. Kortüm

Reflectance Spectroscopy

Principles, Methods, Applications
Translator from the German: J. E. Lohr
160 figures. VI, 366 pages. 1969
Contents: Introduction. – Regular and Diffuse Reflection. – Single and Multiple Scattering. – Phenomenological Theories of Absorption and Scattering of Tightly Packed Particles. – Experimental Testing of the "Kubelka-Munk" Theory. – Experimental Techniques. – Applications. – Reflectance Spectra Obtained by Attenuated Total Reflection. – Appendix: Tables of the Kubelka-Munk-Function. Tables of sin $h^{-1}\chi$; cos $h^{-1}\chi$; cot $h^{-1}\chi$.

NMR
Basic Principles and Progress

Editors: P. Diehl, E. Fluck, R. Kosfeld

Vol. 10:

Van der Waals Forces and Shielding Effects

13 figures, 46 tables. II, 118 pages. 1975

Contents: Van der Waals Forces in NMR Intermolecular Shielding Effects: Historical Development (up to 1961). – Continuum Models. – Pair Interaction Models σ_w. – Other Experimental Proton Data on σ_w. – The Physical Nature of the Field F^2 and of the Associated Excitation Energy. – The Site Factor. – The Repulsion Effect. – The Effects of Higher Order Dispersion Terms. – The Parameters B. – σ_w in Dense Media. – The Temperature Dependence of σ_w. – Factor Analysis. – $^{19}F\, \sigma_w$ Studies. – σ_w of Nuclei other than ^1H and ^{19}F. – Alternate Referencing Systems. – On the Required Molecular Parameters and Physical Constants.

Vol. 11: M. Mehring

High Resolution NMR Spectroscopy in Solids

104 figures. Approx. 240 pages. 1976

Contents: Introduction. – Nuclear Spin Interactions in Solids. – Multiple-Pulse NMR Experiments. – Double Resonance Experiments. – Magnetic Shielding Tensor. – Spin-Lattice Relaxation in Line Narrowing Experiments. – Appendix.

Vol. 12: S. Forsén, B. Lindman

Chlorine, Bromine and Iodine NMR
Physico-Chemical and Biological Applications

72 figures, approx. 45 tables. XIV, 368 pages. 1976

Contents: Introductory Aspects. – Relaxation in Molecules or Ions with Covalently Bonded Halogens. – Shielding Effects in Covalent Halogen Compounds. – Scalar Spin Couplings. – Relaxation of Chloride, Bromide and Iodide Ions. – Shielding of Halide Ions. – Quadrupole Splittings in Liquid Crystals. – Halide Ions in Biological Systems. – Studies of the Perchlorate Ion.

Springer-Verlag Berlin Heidelberg New York